全国餐饮职业教育教学指导委员会重点课题"基于烹饪专业人才培养目标的中高职课程体系与教材开发研究"成果系列教材

餐饮职业教育创新技能型人才培养新形态一体化系列教材

总主编 ◎杨铭铎

烹饪英语

主　编　张会强　崔　健　余松筠
副主编　覃　怀　陈　杰　阎雅瑛　高小芹
编　者　（按姓氏笔画排序）
　　　　杨　格　余松筠　张会强　陈　杰　陈诗尧
　　　　高小芹　黄小梅　崔　健　阎雅瑛　覃　怀

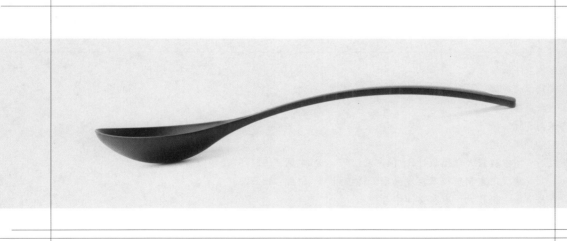

华中科技大学出版社
http://www.hustp.com
中国·武汉

内容简介

本教材为全国餐饮职业教育教学指导委员会重点课题"基于烹饪专业人才培养目标的中高职课程体系与教材开发研究"成果系列教材和餐饮职业教育创新技能型人才培养新形态一体化系列教材。

本教材共10个单元,内容包括烹饪业导论、卫生与安全、厨房设备和工具、调味品和香料、肉类、鱼类、水果和蔬菜、汤、油酥糕点和食谱介绍。本教材用生动形象、通俗易懂的图片、对话、故事等介绍烹饪英语知识,使教材更具可读性,同时配备英文听力材料、教学课件等丰富的数字教学资源。

本教材适用于职业院校烹调工艺与营养、西餐工艺、中西面点工艺及食品营养等专业的学生,也可作为餐饮行业从业人员的烹饪英语入门用书。

图书在版编目(CIP)数据

烹饪英语/张会强,崔健,余松筠主编. —武汉:华中科技大学出版社,2021.1(2023.9 重印)
ISBN 978-7-5680-6874-1

Ⅰ.①烹…　Ⅱ.①张…　②崔…　③余…　Ⅲ.①烹饪-英语-高等职业教育-教材　Ⅳ.①TS972.1

中国版本图书馆 CIP 数据核字(2021)第 013195 号

烹饪英语
Pengren Yingyu

张会强　崔　健　余松筠　主编

策划编辑:汪飒婷	
责任编辑:汪飒婷　马梦雪	
封面设计:廖亚萍	
责任校对:阮　敏	
责任监印:周治超	
出版发行:华中科技大学出版社(中国·武汉)	电话:(027)81321913
武汉市东湖新技术开发区华工科技园	邮编:430223
录　　排:华中科技大学惠友文印中心	
印　　刷:武汉科源印刷设计有限公司	
开　　本:889mm×1194mm　1/16	
印　　张:8.5	
字　　数:326 千字	
版　　次:2023 年 9 月第 1 版第 3 次印刷	
定　　价:49.80 元	

本书若有印装质量问题,请向出版社营销中心调换
全国免费服务热线:400-6679-118　竭诚为您服务
版权所有　侵权必究

全国餐饮职业教育教学指导委员会重点课题
"基于烹饪专业人才培养目标的中高职课程体系与教材开发研究"成果系列教材
餐饮职业教育创新技能型人才培养新形态一体化系列教材

丛书编审委员会

主　任

姜俊贤　全国餐饮职业教育教学指导委员会主任委员、中国烹饪协会会长

执行主任

杨铭铎　教育部职业教育专家组成员、全国餐饮职业教育教学指导委员会副主任委员、中国烹饪协会特邀副会长

副主任

乔　杰　全国餐饮职业教育教学指导委员会副主任委员、中国烹饪协会副会长

黄维兵　全国餐饮职业教育教学指导委员会副主任委员、中国烹饪协会副会长、四川旅游学院原党委书记

贺士榕　全国餐饮职业教育教学指导委员会副主任委员、中国烹饪协会餐饮教育委员会执行副主席、北京市劲松职业高中原校长

王新驰　全国餐饮职业教育教学指导委员会副主任委员、扬州大学旅游烹饪学院原院长

卢　一　中国烹饪协会餐饮教育委员会主席、四川旅游学院校长

张大海　全国餐饮职业教育教学指导委员会秘书长、中国烹饪协会副秘书长

郝维钢　中国烹饪协会餐饮教育委员会副主席、原天津青年职业学院党委书记

石长波　中国烹饪协会餐饮教育委员会副主席、哈尔滨商业大学旅游烹饪学院院长

于干千　中国烹饪协会餐饮教育委员会副主席、普洱学院副院长

陈　健　中国烹饪协会餐饮教育委员会副主席、顺德职业技术学院酒店与旅游管理学院院长

赵学礼　中国烹饪协会餐饮教育委员会副主席、西安商贸旅游技师学院院长

吕雪梅　中国烹饪协会餐饮教育委员会副主席、青岛烹饪职业学校校长

符向军　中国烹饪协会餐饮教育委员会副主席、海南省商业学校校长

薛计勇　中国烹饪协会餐饮教育委员会副主席、中华职业学校副校长

委员（按姓氏笔画排序）

王　劲	常州旅游商贸高等职业技术学校副校长
王文英	太原慈善职业技术学校校长助理
王永强	东营市东营区职业中等专业学校副校长
王吉林	山东省城市服务技师学院院长助理
王建明	青岛酒店管理职业技术学院烹饪学院院长
王辉亚	武汉商学院烹饪与食品工程学院党委书记
邓　谦	珠海市第一中等职业学校副校长
冯玉珠	河北师范大学学前教育学院（旅游系）副院长
师　力	西安桃李旅游烹饪专修学院副院长
吕新河	南京旅游职业学院烹饪与营养学院院长
朱　玉	大连市烹饪中等职业技术专业学校副校长
庄敏琦	厦门工商旅游学校校长、党委书记
刘玉强	辽宁现代服务职业技术学院院长
闫喜霜	北京联合大学餐饮科学研究所所长
孙孟建	黑龙江旅游职业技术学院院长
李　俊	武汉职业技术学院旅游与航空服务学院院长
李　想	四川旅游学院烹饪学院院长
李顺发	郑州商业技师学院副院长
张令文	河南科技学院食品学院副院长
张桂芳	上海市商贸旅游学校副教授
张德成	杭州市西湖职业高级中学校长
陆燕春	广西商业技师学院校长
陈　勇	重庆市商务高级技工学校副校长
陈全宝	长沙财经学校校长
陈运生	新疆职业大学教务处处长
林苏钦	上海旅游高等专科学校酒店与烹饪学院副院长
周立刚	山东银座旅游集团总经理
周洪星	浙江农业商贸职业学院副院长
赵　娟	山西旅游职业学院副院长
赵汝其	佛山市顺德区梁銶琚职业技术学校副校长
侯邦云	云南优邦实业有限公司董事长、云南能源职业技术学院现代服务学院院长
姜　旗	兰州市商业学校校长
聂海英	重庆市旅游学校校长
贾贵龙	深圳航空有限责任公司配餐部经理
诸　杰	天津职业大学旅游管理学院院长
谢　军	长沙商贸旅游职业技术学院湘菜学院院长
潘文艳	吉林工商学院旅游学院院长

网络增值服务

使用说明

欢迎使用华中科技大学出版社医学资源网

1 教师使用流程

（1）登录网址：http://yixue.hustp.com （注册时请选择教师用户）

注册 → 登录 → 完善个人信息 → 等待审核

（2）审核通过后，您可以在网站使用以下功能：

下载教学资源、建立课程、管理学生、布置作业、查询学生学习记录等

2 学员使用流程

（建议学员在PC端完成注册、登录、完善个人信息的操作。）

（1）PC端学员操作步骤

① 登录网址：http://yixue.hustp.com （注册时请选择普通用户）

注册 → 登录 → 完善个人信息

② 查看课程资源：（如有学习码，请在"个人中心—学习码验证"中先通过验证，再进行操作。）

首页课程 → 课程详情页（选择课程）→ 查看课程资源

（2）手机端扫码操作步骤

手机扫码 → 登录 → 查看数字资源
        → 注册

开展餐饮教学研究　　加快餐饮人才培养

餐饮业是第三产业重要组成部分,改革开放40多年来,随着人们生活水平的提高,作为传统服务性行业,餐饮业对刺激消费需求、推动经济增长发挥了重要作用,在扩大内需、繁荣市场、吸纳就业和提高人民生活质量等方面都做出了积极贡献。就经济贡献而言,2018年,全国餐饮收入42716亿元,首次超过4万亿元,同比增长9.5%,餐饮市场增幅高于社会消费品零售总额增幅0.5个百分点;全国餐饮收入占社会消费品零售总额的比重持续上升,由上年的10.8%增至11.2%;对社会消费品零售总额增长贡献率为20.9%,比上年大幅上涨9.6个百分点;强劲拉动社会消费品零售总额增长了1.9个百分点。中国共产党第十九次全国代表大会(简称党的十九大)吹响了全面建成小康社会的号角,作为人民基本需求的饮食生活,餐饮业的发展好坏,不仅关系到能否在扩内需、促消费、稳增长、惠民生方面发挥市场主体的重要作用,而且关系到能否满足人民对美好生活的向往、实现全面建成小康社会的目标。

一个产业的发展,离不开人才支撑。科教兴国、人才强国是我国发展的关键战略。餐饮业的发展同样需要科教兴业、人才强业。经过60多年特别是改革开放40多年来的大发展,目前烹饪教育在办学层次上形成了中职、高职、本科、硕士、博士五个办学层次;在办学类型上形成了烹饪职业技术教育、烹饪职业技术师范教育、烹饪学科教育三个办学类型;在学校设置上形成了中等职业学校、高等职业学校、高等师范院校、普通高等学校的办学格局。

我从全聚德董事长的岗位到担任中国烹饪协会会长、全国餐饮职业教育教学指导委员会主任委员后,更加关注烹饪教育。在到烹饪院校考察时发现,中职、高职、本科师范专业都开设了烹饪技术课,然而在烹饪教育内容上没有明显区别,层次界限模糊,中职、高职、本科烹饪课程设置重复,拉不开档次。各层次烹饪院校人才培养目标到底有哪些区别?在一次全国餐饮职业教育教学指导委员会和中国烹饪协会餐饮教育委员会的会议上,我向在我国从事餐饮烹饪教育时间很久的资深烹饪教育专家杨铭铎教授提出了这一问题。为此,杨铭铎教授研究之后写出了《不同层次烹饪专业培养目标分析》《我国现代烹饪教育体系的构建》,这两篇论文回答了我的问题。这两篇论文分别刊登在《美食研究》和《中国职业技术教育》上,并收录在中国烹饪协会主编的《中国餐饮产业发展报告》之中。我欣喜地看到,杨铭铎教授从烹饪专业属性、学科建设、课程结构、中高职衔接、课程体系、课程开发、校企合作、教师队伍建设等方面进行研究并提出了建设性意见,对烹饪教育发展具有重要指导意义。

杨铭铎教授不仅在理论上探讨烹饪教育问题,而且在实践上积极探索。2018年在全国餐饮职业教育教学指导委员会立项重点课题"基于烹饪专业人才培养目标的中高职课程体系与教材开发研究"(CYHZWZD201810)。该课题以培养目标为切入点,明晰烹饪专业人才培养规格;以职业技能为结合点,确保烹饪人才与社会职业有效对接;以课程体系为关键点,通过课程结构与课程标准精准实现培养目标;以教材开发为落脚点,开发教学过程与生产过程对接的、中高职衔接的两套烹饪专业课程系列教材。这一课题的创新点在于:研究与编写相结合,中职与高职相同步,学生用教材与教师用参考书相联系,资深餐饮专家领衔任总主编与全国排名前列的大学出版社相协作,编写出的中职、高职系列烹饪专业教材,解决了烹饪专业文化基础课程与职业技能课程脱节、专业理论课程设置重复、烹饪技能课交叉、职业技能倒挂、教材内容拉不开层次等问题,是国务院《国家职业教育改革实施方案》提出的完善教育教学相关标准中的持续更新并推进专业教学标准、课程标准建设和在职业院校落地实施这一要求在烹饪职业教育专业的具体举措。基于此,我代表中国烹饪协会、全国餐饮职业教育教学指导委员会向全国烹饪院校和餐饮行业推荐这两套烹饪专业教材。

习近平总书记在党的十九大报告中将"两个一百年"奋斗目标调整表述为:到建党一百年时,全面建成小康社会;到新中国成立一百年时,全面建成社会主义现代化强国。经济社会的发展,必然带来餐饮业的繁荣,迫切需要培养更多更优的餐饮烹饪人才,要求餐饮烹饪教育工作者提出更接地气的教学和科研成果。杨铭铎教授的研究成果,为中国烹饪技术教育研究开了个好头。让我们餐饮烹饪教育工作者与餐饮企业家携起手来,为培养千千万万优秀的烹饪人才、推动餐饮业又好又快的发展,为把我国建成富强、民主、文明、和谐、美丽的社会主义现代化强国增添力量。

姜俊贤

全国餐饮职业教育教学指导委员会主任委员
中国烹饪协会会长

出版说明

《国家中长期教育改革和发展规划纲要(2010—2020年)》及《国务院办公厅关于深化产教融合的若干意见(国办发〔2017〕95号)》等文件指出：职业教育到2020年要形成适应经济发展方式的转变和产业结构调整的要求，体现终身教育理念，中等和高等职业教育协调发展的现代教育体系，满足经济社会对高素质劳动者和技能型人才的需要。2019年1月，国务院印发的《国家职业教育改革实施方案》中更是明确提出了提高中等职业教育发展水平、推进高等职业教育高质量发展的要求及完善高层次应用型人才培养体系的要求；为了适应"互联网＋职业教育"发展需求，运用现代信息技术改进教学方式方法，对教学教材的信息化建设，应配套开发信息化资源。

随着社会经济的迅速发展和国际化交流的逐渐深入，烹饪行业面临新的挑战和机遇，这就对新时代烹饪职业教育提出了新的要求。为了促进教育链、人才链与产业链、创新链有机衔接，加强技术技能积累，以增强学生核心素养、技术技能水平和可持续发展能力为重点，对接最新行业、职业标准和岗位规范，优化专业课程结构，适应信息技术发展和产业升级情况，更新教学内容，在基于全国餐饮职业教育教学指导委员会2018年度重点课题"基于烹饪专业人才培养目标的中高职课程体系与教材开发研究"(CYHZWZD201810)的基础上，华中科技大学出版社在全国餐饮职业教育教学指导委员会副主任委员杨铭铎教授的指导下，在认真、广泛调研和专家推荐的基础上，组织了全国90余所烹饪专业院校及单位，遴选了近300位经验丰富的教师和优秀行业、企业人才，共同编写了本套餐饮职业教育创新技能型人才培养新形态一体化系列教材、全国餐饮职业教育教学指导委员会重点课题"基于烹饪专业人才培养目标的中高职课程体系与教材开发研究"成果系列教材。

本套教材力争契合烹饪专业人才培养的灵活性、适应性和针对性，符合岗位对烹饪专业人才知识、技能、能力和素质的需求。本套教材有以下编写特点：

1. 权威指导，基于科研　本套教材以全国餐饮职业教育教学指导委员会的重点课题为基础，由国内餐饮职业教育教学和实践经验丰富的专家指导，将研究成果适度、合理落脚于教材中。

2. 理实一体，强化技能　遵循以工作过程为导向的原则，明确工作任务，并在此基础上将与技能和工作任务集成的理论知识加以融合，使得学生在实际工作环境中，将知识和技能协调配合。

3. 贴近岗位，注重实践　按照现代烹饪岗位的能力要求，对接现代烹饪行业和企业的职业技能标准,将学历证书和若干职业技能等级证书("1+X"证书)内容相结合,融入新技术、新工艺、新规范、新要求,培养职业素养、专业知识和职业技能,提高学生应对实际工作的能力。

4. 编排新颖,版式灵活　注重教材表现形式的新颖性,文字叙述符合行业习惯,表达力求通俗、易懂,版面编排力求图文并茂、版式灵活,以激发学生的学习兴趣。

5. 纸质数字,融合发展　在新形势媒体融合发展的背景下,将传统纸质教材和我社数字资源平台融合,开发信息化资源,打造成一套纸数融合的新形态一体化教材。

本系列教材得到了全国餐饮职业教育教学指导委员会和各院校、企业的大力支持和高度关注,它将为新时期餐饮职业教育做出应有的贡献,具有推动烹饪职业教育教学改革的实践价值。我们衷心希望本套教材能在相关课程的教学中发挥积极作用,并得到广大读者的青睐。我们也相信本套教材在使用过程中,通过教学实践的检验和实际问题的解决,能不断得到改进、完善和提高。

前言

2019年1月，国务院下发了《国家职业教育改革实施方案》，提出了"推进高等职业教育高质量发展"和"完善高层次应用型人才培养体系"等要求，本教材就是按照方案的要求进行设计和编撰的。

"烹饪英语"是高职高专院校烹饪行业内烹调工艺与营养专业、西餐工艺专业、中西面点工艺专业及相关专业的一门专业核心必修课。本课程通过介绍中西餐饮文化、厨房规则、厨房设施设备、各类烹饪原料名称、各类烹饪方法、中西菜品的翻译原则等任务模块的烹饪英语知识，使学生能够系统全面地掌握烹饪英语的核心词汇和表达方式，具备运用烹饪英语进行沟通交流的能力，为将来从事国际烹饪与餐饮工作打下基础。

随着经济的发展和对外交流的日益增多，人们对餐饮服务从业者的要求不断提高，尤其是对专业英语能力提出了更高的要求。本教材以职业活动为载体，以素质培养为基础，以职业技能标准为结合点，以培养学生英语应用能力为关键点，以学生为主体，以训练为手段，突出语言应用能力目标，设计出知识、技能、素质一体化的职业课程。本教材注意和中职烹饪专业英语相衔接，始终把创新思想贯穿于教材的编写中，以烹饪英语各方面的知识为基础，以交际能力为目标；对烹饪知识的专业性、技能性和应用性进行创新管理，有意识地培养学生的烹饪专业英语交际能力；以提高学生职业素养为切入点，以培养学生知识目标为结合点，以培养学生适应岗位需求的交际能力为关键点，要求学生具备十二种职业能力，其中之一就是广泛进行应用英语交流，具备较高的英语表达能力和沟通能力。本教材整合线上、线下的立体化教学资源，依托数字化平台，结合高职院校专业优势，进行资源整合，是开发行业发展需求的数字化、立体化应用型行业英语教材。英语是学生学习行业知识的媒介和工具，"烹饪英语"的学习目的就是提高烹饪人才在国际化环境下的交际能力。

本教材由青岛酒店管理职业技术学院张会强担任第一主编，黑龙江旅游职业技术学院崔健和武汉商学院余松筠担任第二、第三主编。张会强负责框架设计、审稿和定稿工作。具体章节分工如下：云南能源职业技术学院覃怀负责 Unit 1 的编写工作；崔健负责 Unit 2 和 Unit 4 的编写工作；余松筠负责 Unit 3 和 Unit 7 的编写工作；张会强负责 Unit 5 和 Unit 6 的编写工作；郑州商业技师学院黄小梅负责 Unit 8 的编写工作；淄博市技师学院陈诗尧负责 Unit 9 的编写工作；长沙商贸旅游职业技术学院陈杰负责 Unit 10 的编写工作；酒泉职业技术学院阎雅瑛、三峡旅游职业技术学院高小芹负责听力材料等数字资源的制作；北京2022年冬奥会和冬残奥会组织委员会杨格负责资料收集和数字化处理等工作。这些老师多数是从事一线教学工作多年的专任教师,具有丰富的教学经验和教改经验,在编写过程中把教学的感悟融入教材之中。

  本教材的编写得到了华中科技大学出版社及青岛酒店管理职业技术学院各位领导的热情帮助和大力支持,在此表示深深的谢意!同时,本教材的编写也参阅了国内外大量的专著和书籍,在此对借鉴书刊、资料的作者深表谢意。

  学力不逮之处,颛此就正于方家。以期在今后的教学中,得到改进和提高。

编者

目录

Unit 1 Introduction to Cooking Industry 1

Part A Introduction to Chinese Food Culture and Major Chinese Cuisines 1
Part B Introduction to Western Food Culture and Knowledge of Western Food 5

Unit 2 Kitchen Introduction 10

Part A Kitchen Positions and Rules 10
Part B Kitchen Hygiene 13
Part C Kitchen Safety 16

Unit 3 Tools and Equipment 20

Part A Hand Tools 20
Part B Pots and Pans 23
Part C Knives 26
Part D Cooking Equipment 28

Unit 4 Condiments and Spices 34

Part A Condiments 34
Part B Spices 38

Unit 5 Meat 43

Part A Understanding Meat of Domestic Animals and Wild Animals 43
Part B Poultry 49

Unit 6 Fish   55

Part A Understanding Fish and Shellfish	55
Part B Shellfish	58
Part C How to Cook a Seafood Dish	61

Unit 7 Fruits and Vegetables   66

Part A Fruits	66
Part B Vegetables	68
Part C Processing Methods of Fruits and Vegetables	71

Unit 8 Soup   75

Part A Classification of Soup	75
Part B French Onion Soup	78
Part C Chicken Soup	80
Part D Eel Soup	81

Unit 9 Pastry   84

Part A Chinese-style Pastry	84
Part B Western-style Pastry	87
Part C Mounting Decoration	91

Unit 10 Menu Recommendation   95

Part A Starters	95
Part B Main Course	98

Appendix	**I** **Answers**	103
Appendix	**II** **Vocabulary**	113

Unit 1

Introduction to Cooking Industry

扫码看课件

Part A Introduction to Chinese Food Culture and Major Chinese Cuisines

  Learning Goals

You will be able to:
1. know the characteristics of Chinese food culture;
2. know the differences between the eight cuisines.

Vocabulary Assistance

flavor ['fleɪvə]   n. 味道,特点,特色
                   v. 给……调味,给……增添风味
cuisine [kwɪ'ziːn]   n. 烹饪,风味菜肴
mouthwatering ['maʊθwɔːtərɪŋ]   adj. 令人垂涎的
greasy ['ɡriːsi]   adj. 油腻的,谄媚的,多油的
shallot [ʃə'lɒt]   n. 葱,红葱头
garlic ['ɡɑːrlɪk]   n. 大蒜,蒜头
pungent ['pʌndʒənt]   adj. 辛辣的,刺激性的,说穿了,一针见血的
soybean ['sɔɪbiːn]   n. 大豆,黄豆
aroma [ə'roʊmə]   n. 芳香,香味
prolific [prə'lɪfɪk]   adj. (艺术家、作家等)多产的,众多的,富饶的,(植物、动物等)丰硕的
chili ['tʃɪli]   n. 红辣椒
pepper ['pepə(r)]   n. 甜椒,辣椒,胡椒粉
                    v. 在……上撒胡椒粉,使布满,连续击打

ginger ['dʒɪndʒə(r)]   n. 姜,生姜,姜黄色
                       v. 使活跃,使有活力
                       adj. 姜黄色的
chef [ʃef]   n. 厨师,大师傅
savory ['seɪvəri]   n. (烹调用的)香薄荷
                    adj. 好吃的,咸味的
pickle ['pɪkl]   v. 腌渍(泡菜等)
                 n. 腌菜,泡菜,腌制食品
palate ['pælət]   n. 腭,上腭,味觉,品尝力
ingredient [ɪn'ɡriːdiənt]   n. (混合物的)组成部分,(烹调的)原料,(成功的)要素,因素
mellow ['meloʊ]   adj. (瓜、果等)成熟的,(酒)芳醇的,(颜色或声音)柔和的,老练的
                  v. (使)成熟,变柔和,使芳醇
fragrance ['freɪɡrəns]   n. 芳香,芬芳,香气,香水
braise [breɪz]   v. 炖,焖

Text

In China, it is said that food is the paramount necessity of the common people. Culinary culture is a very important part of Chinese culture. Because China has a vast territory, a large population, and great differences in the natural geographical environment of various regions, people's food culture is also different. Usually, Chinese cuisine is divided into eight major cuisines, each of which has its own characteristics and is integrated into each other, which together constitute a rich and colorful Chinese food culture.

Shandong cuisine

As one of the eight major cuisines, Shandong cuisine, composed of Jinan cuisine and Jiaodong cuisine, is popular in northern China. Shandong cuisine pays attention to the freshness and crispness of the taste. In the process of making Shandong cuisine, garlic and onion are widely used as condiments. In addition, another feature of Shandong cuisine is the emphasis on soup matching, usually with clear, fresh thin soup and rich flavor of cream soup. Jinan cuisine is mainly fried, while Jiaodong cuisine is famous for its fresh and light seafood cooking.

Sichuan cuisine

Sichuan cuisine is one of the most popular Chinese dishes in the world. Because Sichuan cuisine is the production of a large number of peppers, producing a strong stimulating taste, so Sichuan cuisine is characterized by hot and numbing taste. In addition, garlic, ginger and fermented soybeans are also used extensively during cooking. As to cooking techniques, it usually includes frying, oil-free frying, pickling and stewing.

Once there was a saying that if a man had not eaten Sichuan food, he could not prove that he had been to China.

Guangdong cuisine

Cantonese food originates from Guangdong, the southernmost province in China. The majority of overseas Chinese people are from Guangdong (Canton), so Cantonese is perhaps the most widely available Chinese regional cuisine outside of China.

Cantonese are known to have an adventurous palate, able to eat many different kinds of meat and vegetables. Cantonese food is one of the most diverse and richest cuisines in China. Many vegetables originate from other parts of the world. It doesn't use many spices, bringing out the natural flavor of the vegetables and meat. Tasting clear, light, crisp and fresh, Guangdong cuisine, familiar to westerners, usually chooses raptors and beasts to produce originative dishes. Its basic cooking techniques include roasting, stir-frying, deep-frying, braising, stewing and steaming. Among them steaming and stir-frying are more commonly applied to preserve the natural flavor. Guangdong chefs also pay much attention to the artistic presentation of dishes.

Fujian cuisine

Among the eight major cuisines, Fujian cuisine is distinctive. Usually, Fujian cuisine contains three kinds of local cuisines, which are Fuzhou cuisine, Quanzhou cuisine and Xiamen cuisine. Fujian cuisine is usually mainly seafood; paying attention to beautiful color collocation, in the taste of being sweet and sour, salty taste is famous.

Jiangsu cuisine

Jiangsu cuisine is also known as Huaiyang cuisine because it originated in the area flowing through the Huai River. Aquatics as the main ingredients, it stresses the freshness of materials. Its carving techniques are delicate, of which the melon carving technique is especially well known. Cooking techniques consist of stewing, braising, roasting, simmering, etc. The flavor of Huaiyang cuisine is light, fresh and sweet and with delicate elegance. Jiangsu cuisine is well known for its careful selection of ingredients, its meticulous preparation methodology, and its not-too-spicy and not-too-bland taste. Since the seasons vary in Jiangsu, the cuisine also varies throughout the year. If the flavor is strong, it isn't too heavy; if light, not too bland.

Zhejiang cuisine

The favorite Zhejiang cuisine is made up of Hangzhou cuisine, Ningbo cuisine and Shaoxing cuisine. Zhejiang cuisine is not greasy and is famous for its freshness, softness and smoothness. Hangzhou cuisine is the most famous one among the three.

Hu'nan cuisine

Another dish popular with Chinese people is Hu'nan cuisine. Hu'nan cuisine consists of local cuisines of Xiangjiang Region, Dongting Lake and Xiangxi Coteau. It characterizes itself by thick and pungent flavor. Chili, pepper and shallot are usually necessaries in this division.

Anhui cuisine

Anhui cuisine chefs focus much more attention on the temperature in cooking and are good at braising and stewing. Often hams will be added to improve taste and sugar candy added to gain.

Guest: Excuse me, I'd like to try some Chinese food.
Waiter: We serve excellent Chinese food. Which style do you prefer?
Guest: I know nothing about Chinese food. Could you give me some suggestions?
Waiter: It's divided into 8 big cuisines such as Cantonese food, Shandong food, Sichuan food, etc.
Guest: Is there any difference?
Waiter: Yes, Cantonese food is lighter while Shandong food is heavier and spicy.
Guest: How about Sichuan food?
Waiter: Most Sichuan dishes are spicy and hot. They taste differently.
Guest: Oh, really. I like hot food. So what is your recommendation for me?
Waiter: I think Mapo bean curd and shredded meat in chili sauce are quite special and delicious. We have a Sichuan food dining room. May I suggest you to go there? It's on the third floor.
Guest: Okay, I will try. Thank you.

Activity I What cuisines can you see in the pictures? Please write them down, then dictate their characters.

烹饪英语

1. _____ 2. _____

3. _____ 4. _____

5. _____ 6. _____

7. _____ 8. _____

Unit 1 Introduction to Cooking Industry

Activity Ⅱ Listen to the following passage and fill in the blanks.

In China, the festival celebration without delicacies, as if food has become an integral part of Christmas. It also indicates the diet will surely become the Chinese culture. It is an 1._____ part in the festival.

China is probably "eat" culture countries with the longest. We all aspects of life "to eat" penetrates the theme. "Eat" culture can produce so much influence, of course, and its close contact with our life is not divided, but more important, it can follow the pace of The Times, advancing with The Times, the constant 2._____ innovation.

Mention of food culture, we first thought of "eat". Although "eat" culture is the diet culture as a 3._____ part, but to a great extent, it can 4._____ our food culture.

China's diet in the world is 5._____, Chinese food color, fragrance, taste, shape of praise. Having Chinese place is due to a Chinese restaurant, Chinese food can be said to be "feed" the world. This phenomenon early in this century at first was the forerunner of the 6._____ of Sun yat-sen keenly observed. Dr. Sun yat-sen said: "I modern Chinese civilization evolution, everything fell into later, but a diet of progress, are still not for all countries." Dr. Sun yat-sen's discussion is quite correct, but in fact it such as Dr. Sun yat-sen says, the Chinese 7._____ in modern times, the impact of western civilization plate the goal, would ask a 8._____ of insight's pulse. China, however, the restaurant is their forebears, incredibly to log on to Europe and America, all over the world, invincible, so far the world almost every corner all have Chinese restaurant.

Activity Ⅲ Translate a passage from Chinese into English.

在西方人心目中，和中国联系最为密切的基本食物是大米。长期以来，大米在中国人的饮食中占据很重要的地位，以至于有谚语说"巧妇难为无米之炊"。中国南方大多数地区种植水稻，人们通常以大米为食；而华北大部分地区因为过于寒冷或过于干燥，无法种植水稻，主要作物是小麦。在中国，有些人用面粉做面包，但大多数人用面粉做馒头和面条。

Tips

The culture of China is extensive and profound, and the Chinese culture of food is rich and varied. Learning food culture also has many advantages.

So, we should inherit the fine traditional of Chinese culture. Learning from it and make ourselves grow better. In the meantime, we can rich Chinese culture through our strength. Let the world see the Chinese culture.

Part B Introduction to Western Food Culture and Knowledge of Western Food

Learning Goals

You will be able to:

1. know the classification of western food;
2. tell the name of the common food of different western countries;
3. know how to eat western food.

Vocabulary Assistance

olive ['ɒlɪv]　n. 橄榄，油橄榄，橄榄树，橄榄色
　　　　　　　adj. 橄榄绿的，黄褐色的，淡褐色的
bacon ['beɪkən]　n. 培根，熏猪肉，咸猪肉，腊肉
porridge ['pɒrɪdʒ]　n. 粥，麦片粥，稀饭
butter ['bʌtə(r)]　n. 黄油，黄油状的食品
　　　　　　　　v. 抹黄油于……上，用黄油煎食物，讨好
cereal ['sɪəriəl]　n. 谷类植物，谷物，粮食，谷类食品
dessert [dɪ'zɜːt]　n. 餐后甜食，甜点
vegetarian [ˌvedʒə'teəriən]　n. 素食者，食草动物
　　　　　　　　　　　adj. 素食者的，素菜的

exquisite [ɪk'skwɪzɪt]　adj. 精致的，细腻的，优美的，剧烈的
pasta ['pæstə]　n. 面团（用以制意大利通心粉、细面条等），意大利面食
sausage ['sɒsɪdʒ]　n. 香肠，腊肠
salad ['sæləd]　n. 沙拉，色拉
risotto [rɪ'zɒtəʊ]　n. 意大利肉汁烩饭
vinegar ['vɪnɪɡə(r)]　n. 醋
essence ['esns]　n. 本质，实质，精华，精髓，香精
immigrant ['ɪmɪɡrənt]　n. 移民，侨民，从异地移入的动物（植物）
ludicrous ['luːdɪkrəs]　adj. 可笑的，荒唐的
spoon [spuːn]　n. 勺，匙，调羹，一匙的量，匙状物，匙桨
　　　　　　v. 用汤匙舀取，轻轻向上击

 Text

　　In the long history, westerners form a diet culture with their own unique cultural characteristics through the accumulation and creation of diet. With the continuous improvement of the influence of the West on the world, western cuisine also has an important influence on the diet of the world today.

　　In England, people always have some bread and milk for breakfast on weekdays which is always in a hurry. But on weekends, breakfast is a big feast with bacon, porridge, coffee, eggs, cakes, and so on. As for lunch, it's always so simple. People usually have a sandwich or a hot dog. Some people also eat in a fast-food restaurant nearby. This meal won't last very long. Dinner is the biggest meal with potato, beef, chicken, soup and some desserts. At this meal, people usually sit around the table and talk about their daily events. This is the best time for family gathering. Cheese is the most popular food in European countries and the United States. According to different dishes and wine, different cheese is tied in. The simplest cheese is to put a certain amount of all kinds of cheese on a plate, called cheese dish.

　　Jewish usually eat apples with honey on New Year's Day, in order to celebrate the happiest New Year.

　　Western food loves the production of excellence and the pursuit of perfection, attaching importance to food, tableware and dining environment. Among them, the famous western diet includes Italian food, French food, American food, German food, Russian food, and so on. Western diet is characterized by meat-based, vegetarian supplemented with wine, coffee and black tea.

The main food in Britain are meat, eggs, pasta, milk and dairy products. The British ones are light in taste, exquisite in cuisine and varied in variety. British dinner is the most abundant of the day, usually drinking a variety of wine, after the meal to drink sweet wine, coffee or black tea.

German staple food is wheat, potatoes, and so on. Bread is German favorite food, of course, and Germans also like to eat cheese, sausage, with raw vegetables salad and fruits.

French cuisine in general includes these aspects: bread, pastry, cold food, cooked food, meat products, cheese and wine. These are indispensable elements of the French diet, and the most proud of them are wine, bread and cheese.

We are familiar with Italian food pasta, pizza, risotto, vinegar and Italian ice cream, coffee, and so on. But Italian food is not just about that. On the contrary, Italian cuisine is very rich, with different regions and different towns. The difference between Italian cuisine and other countries cuisines is that it is rich in ingredients and can be made at will, so that its essence lies in self-expression.

American diet is not exquisite, the pursuit of quick and convenient, nor luxury, more popular. Because the United States is an immigrant country, the cooking style and characteristics brought by the new immigrants make the table of American families and restaurants rich and diversified food. Americans must drink three meals a day, usually with wine, beer or milk, and spirits are usually consumed in bars. Americans like dessert, tea or coffee after meals.

American eating is funny. They eat almost everything with a fork, and it appears that holding a knife in one's right hand longer than a few seconds is considered to be against good table manners. The system is that if it is absolutely necessary to use a knife, people take the fork in their left hand, and cut off a piece of meat or whatever it is in the normal manner. Then they put the knife down, transfer the fork to their right hand, and only then do they transport the food to their mouth. This is clearly ludicrous, but it is considered good manners.

There are several results of this system. First, if it is not absolutely necessary to use a knife, Americans don't use one, because obviously this greatly complicates things, and you will therefore see them trying to cut things like potatoes, fish and even bacon with a fork. Second, towards the end of a course, since only one implement is being used, food has to be chased around the plate with the fork, and for the last mouthful the thumb has to be used to keep the food in place, although one is not supposed to do this. Third, tables are generally laid with one knife and two forks, the outside fork being for the salad. There is no need for the foreign visitors to follow the American system and try to eat the salad with only a fork, but if you do use your knife, remember to save it for the meat course. Even desserts are eaten with a fork if at all possible, and the spoon you see by your dessert is meant to be for coffee.

A: May I take your order?

B: Do you have beef steak?

A: Yes, we also have salad, pizza, ice cream, pudding, spaghetti.

B: I would like beef steak with salad.

A: How do you like it cooked?

B: Medium well.

外教有声

C: I would like pizza with ice cream.
D: I would like spaghetti.
A: Beef steak with salad, pizza with ice cream, spaghetti.
B/C/D: That's right.

Exercises

Activity Ⅰ What is your favorite food? Please draw a tick under the picture, and dictate its characters.

1._____

2._____

3._____

4._____

5._____

6._____

7._____   8._____

Activity II Choose an option that fits you.

1. How do you like your egg cooked? _____
   A. scramble   B. sunny side up   C. ease over   D. boil
2. How do you like your steak cooked? _____
   A. well done   B. medium well   C. medium   D. rare

Activity III Translate the following passage into Chinese.

Do not eat food with your fingers unless you are eating food customarily eaten with fingers, such as bread, French fries, chicken wings, pizza, etc.

The fork may be used either in the American (use the fork in your left hand while cutting; switch to right hand to pick up and eat a piece) or the continental (fork remains in the left hand) fashion, either is now acceptable. The fork is used to convey any solid food to the mouth.

When you have finished eating soup, the spoon should be placed to the side of the saucer, not left in the bowl.

Keep your napkin on your lap. At more formal occasions all diners will wait to place their napkins on their laps until the host or hostess places his or her napkin on his or her lap. When eating barbecue or some other messy foods such as cracked crab, a bib napkin may be provided for and used by adults. Usually these foods are also eaten by hand, and wet wipes or paper napkins should be used to clean the hands.

Unit 2

Kitchen Introduction

Part A   Kitchen Positions and Rules

 Learning Goals

You will be able to:
1. know main positions of kitchen;
2. know the duties of kitchen position;
3. know basic kitchen rules.

 Vocabulary Assistance

chef [ʃef]　n. 厨师,主厨,厨师长
sous chef　副厨师长,副主厨
mysterious [mɪˈstɪəriəs]　adj. 神秘的,不可思议的
derived from　来源于,源自
executive chef　行政总厨,行政主厨,厨师长
second in command　副指挥,第二把手
hierarchical [ˌhaɪəˈrɑːkɪkl]　adj. 分层的,等级体系的
brigade [brɪˈɡeɪd]　n. 团队,大军
delineate [dɪˈlɪnieɪt]　v. 描绘,描写,画……的轮廓
in most cases　大部分情况下
application [ˌæplɪˈkeɪʃn]　n. 应用,申请
an integral part　不可分割的部分
management [ˈmænɪdʒmənt]　n. 管理,管理人员
culinary [ˈkʌlɪnəri]　adj. 烹饪的,烹饪用的
hallmark [ˈhɔːlmɑːk]　n. 特点,品质证明
entry-level　入门的,初级的
designation [ˌdezɪɡˈneɪʃn]　n. 指定
precise [prɪˈsaɪs]　adj. 明确的

operation [ˌɒpəˈreɪʃn]　n. 操作,经营
professional [prəˈfeʃnl]　adj. 专业的,职业的
cook [kʊk]　n. 厨师
commis [ˈkɒmi]　n. 初级厨师,厨师助理
pastry chef　面点厨师长
sommelier [ˈsɒməljeɪ]　n.（法）酒侍,品酒师
prep cook　备餐厨师
larder chef　负责烹饪肉类的厨师长
vegetable chef　负责烹饪蔬菜的厨师长
fish cook　负责烹饪鱼类的厨师
butcher [ˈbʊtʃə(r)]　n. 屠夫
pantryman [ˈpæntrɪmən]　n. 配餐员
potman　擦洗锅的人
relief cook　替班厨师
master chef　厨艺大师
personal chef　私人厨师
baker [ˈbeɪkə(r)]　n. 面包师,烘焙师
sauce chef　调汁厨师
fry cook　负责煎炸的厨师
catering [ˈkeɪtərɪŋ]　n.（会议或社交活动的）饮食服务,酒席承办

Life of a Sous Chef

What is a sous chef? The mysterious sounding name is derived from the French word meaning "under". This kitchen works closely with the executive chef, as second in command, in the classic hierarchical kitchen structure known as "brigade". The brigade system formally delineates the position of each kitchen worker, and while its strict application depends on the type and size of the kitchen, in most cases a sous chef will be present, as an integral part of the management team.

For culinary students, obtaining a sous chef position is often eyed as a hallmark level of achievement. On-the-job experience is essential for any kitchen worker who wants to climb the ladder, so sous chef is generally not considered to be an entry-level position. If you are a culinary student whose ultimate goal is to be an executive chef, focus your efforts on mastering each kitchen "station", with an eye toward advancement.

As a sous chef working in a large kitchen or institution, you may be one of several staffers who hold the position. One way or another, the sous chef designation bridges the professional gap between line cooks and the executive chef.

In practice, many kitchens keep long daily hours of operation. Hotels, resorts, cruise ships and other hospitality outlets provide breakfast, lunch and dinner. That's a lot of hours to account for, so sous chefs often work opposite their chefs, to provide management coverage during the chef's off-time.

As second in command, your job as sous chef can be an opportunity waiting to happen. As chefs move on, or are promoted to food and beverage directorships and general manager jobs, sous chefs are perfectly positioned to step in to executive chef roles.

外教有声

Peter: Hi, my name is Peter Johnson. I'm the sous chef of inn fan restaurant. Welcome to join us.

Thomas: Nice to meet you too, Mr. Johnson. My name is Thomas King. I'm just graduate from the cooking vocational school. So what should I do as a commis?

Peter: Oh, you should learn how to help other cooks when they are cooking. I'm responsible for staff training and cooking. If you have any questions you can come to me for help.

Thomas: Thanks a lot. It's very kind of you. I'll try my best to do it.

Peter: You will work in the pastry section. The woman over there is the pastry chef. She is in charge of making pastry. You are going to work with her.

Thomas: Oh, I got it.

Exercises

Activity Ⅰ Find the answers.

Where is the	executive chef
	pastry chef
	sommelier
	prep cook
	larder chef
	vegetable chef
	fish cook
	butcher

He is in the	chef's office
	pastry section
	beverage cooler
	hot kitchen
	cold kitchen
	vegetable preparation section
	fish section
	butchery

Activity Ⅱ Do you know the names of the kitchen positions? Match the following sentences with the positions.

> Pantry man   Butcher   Potman   Relief cook   Executive chef
> Breakfast cook   Larder chef   Roast cook   Pastry chef   Commis

1. _____ The person who can relieve everyone.
2. _____ The person who is the head of the chefs and works in the chef's office.
3. _____ The person who is in charge of all kinds of meat.
4. _____ The person who is responsible for the pantry.
5. _____ The person who cuts the meat and slaughters the animals.
6. _____ The person who makes the breakfast.
7. _____ The person who washes pans and pots.
8. _____ The person who is in the hot kitchen to roast meat.
9. _____ The person who is in charge of the pastry production process.
10. _____ The person who allocated to a section to assist the chef.

Focus on Language

Listen to the tape and write the chef's positions.

1. _____ This person is responsible for all aspects of the kitchen, from menu to management. They make all decisions regarding how food is prepared and presented.

2. _____ This chef is considered the assistant to the executive chef and is responsible for filling in as needed and serving as the second in command in a kitchen.

3. _____ This person is very knowledgeable about wine, and is responsible for wine service, pairing and storage. It is their job to train team members and suggest wine to compliment food.

4. _____ This chef is one who prepares specifically for a client in their home. Some prepare meals in bulk to be frozen and eaten over a period of time, and some may be responsible for cooking meals for large groups for certain events.

5. _____ This chef is a prestigious title, and one receives it after achieving the highest level certification. It requires extensive training in various programs. There are only a small number in the U.S.

6. _____ This chef is experienced in dessert and baked good and often works in large kitchens, such as in hotels and restaurants. In some settings, this chef is also responsible for creating the dessert menu.

7. _____ This person is responsible for preparing ingredients and other materials for the chefs. Some of their duties might include cleaning and chopping vegetables or preparing soup and dressings.

8. _____ People in this business prepare and serve food at offsite locations, such as homes, hotels, or other facilities. Some examples are weddings, receptions, or other private events.

9. _____ This position is similar to a pastry chef, and in some settings, the two names are used interchangeably. They are responsible for baking cakes, bread and other goods.

10. _____ This chef is responsible for fried menu items and duties related to keeping fryers and their oils in operational condition. Onion rings, calamari, chicken cordon bleu and a host of other deep-fried delicacies are put forth by this specialist.

11. _____ This chef specializes in sauces and marinades. In most commercial settings, the saucier is responsible for a variety of additional tasks, such as sautéing food and preparing soup and stew.

Part B Kitchen Hygiene

Learning Goals

You will be able to:

1. know some rules of kitchen hygiene;
2. know the importance of keeping food fresh.

Vocabulary Assistance

hygiene ['haɪdʒiːn]　n. 卫生
sanitation [ˌsænɪ'teɪʃn]　n. 公共卫生
contaminated [kən'tæmɪneɪtɪd]　adj. 受污染的,弄脏的
prevent [prɪ'vent]　v. 预防,防止,阻止
stray [streɪ]　adj. 走失的,孤立的
sleeve [sliːv]　n. 袖子
roll up　卷起
snag [snæg]　v. 被绊住,形成障碍
potentially [pə'tenʃəli]　adv. 可能地,潜在地
cross-contamination [ˌkrɔːskənˌtæmɪ'neɪʃn]　n. 交叉污染
bacteria [bæk'tɪəriə]　n. 细菌
stack [stæk]　n. (整齐的)一堆
bleach [bliːtʃ]　n. 漂白剂,消毒剂

regulation [ˌregju'leɪʃn]　n. 管理,规则
disinfect [ˌdɪsɪn'fekt]　v. 给……消毒
policy ['pɒləsi]　n. 政策,方针
procedure [prə'siːdʒə(r)]　n. 程序,手续,步骤
spread [spred]　v. 传播,散布
symptom ['sɪmptəm]　n. 症状
infect [ɪn'fekt]　v. 感染,传染
sterilize ['sterəlaɪz]　v. 消毒,杀菌
dangerous ['deɪndʒərəs]　adj. 危险的
refrigerate [rɪ'frɪdʒəreɪt]　v. 冷藏,冷冻
cough [kɒf]　v. 咳嗽
sneeze [sniːz]　v. 打喷嚏
clean as you go　随手清洁
utensil [juː'tensl]　n. 餐具,炊具
chopping board　案板,切菜板

 **Text**

Basic Hygiene Rules You Should Follow in the Kitchen

　　Personal hygiene is essential to the safety and sanitation of the kitchen. If you are dirty, your kitchen will be dirty and the food could be contaminated. Keep your hair up to prevent any stray hair getting into the food. Wear clothes that fit you well and keep your sleeves rolled up. This is to prevent snagging on anything and potentially pulling something over or dropping something on yourself. Be sure to wash your hands in warm water and soap frequently to prevent cross-contamination of bacteria.

　　When storing food, always follow the FIFO plan: First In, First Out. Newer food is stored behind or on the bottom of the stacks, leaving older food more likely to be used first. Freeze all meat that are not going to be used in a couple of days. Do not allow hot food to cool or cold food to warm before serving.

　　Take a page out of the fast food industry's sanitation regulations and wipe down all hard surfaces after using them. Use hot water to which a little bit of bleach has been added. This will remove dirt from the surfaces and will disinfect those surfaces at the same time. Keeping a clean kitchen is essential to the health of all those eating from that kitchen.

　　Food safety is one of the most important tasks given to a restaurant manager and staff; policies and procedures must be followed in order to ensure that food-borne illnesses are not allowed to spread.

Employees with cough, sneeze or other symptoms should stay home so as not to infect fellow workers.

 Dialogue

Commis: How do microbes (bacteria) get in food?
Chef: Mostly from your hands.
Commis: I should always wash them.
Chef: Yes, everything in the kitchen should be sanitized.
Commis: Right, I'll use an antiseptic to sterilize.
Chef: Great!
Commis: Chef, what foods are most dangerous?
Chef: Protein foods: meat, poultry, eggs, fish, dairy products, etc.
Commis: Does food poisoning have a special smell or a special color?
Chef: No.
Commis: How can I stop food poisoning?
Chef: Refrigerate food below 40 ℉.
Commis: Is there anything else I should know?
Chef: Never leave food outside the refrigerator for more than two hours.

 Exercises

Activity Ⅰ Ask and answer the following questions with a partner, using a complete sentence.

1. How do bacteria get in food? Can you give more reasons?
2. Why should everything in the kitchen be sanitized?
3. What foods are most dangerous? Why?
4. Does food poisoning have a special taste, a special smell or a special color?
5. How can we stop food poisoning?
6. Why should we refrigerate food below 40 ℉?
7. Why should we always keep food in the refrigerator?
8. Can we leave food outside the refrigerator for more than two hours? Why?
9. Do you think it is important for us to keep the kitchen clean? Why?
10. What do you learn from this text?

Activity Ⅱ Do you know how to wash your hands?

soap   lather   wet   scrub   dry   rinse

First, 1._____ your hands with clean running water and use soap. Use warm water if it is available.
Second, 2._____ your hands.
Third, 3._____ hands together to make lather.
Then, 4._____ all surfaces.
Next, 5._____ hands well under running water.
Finally, 6._____ your hands using a paper towel or air dryer.

Focus on Language

🎧 **Listen to the tape and fill in the blanks with the missing words.**

Here Are some Basic Hygiene Rules You Should Follow

To avoid breeding bugs and accumulated germs in your kitchen, you need to 1._____ to ensure it stays a 2._____ for your guest.

Rule 1: Clean kitchen surfaces after every stage of preparing your recipe. Try to 3._____. This may sound a little obsessive, but it's not. Raw meat, poultry, fish, eggs and many other raw foods are the most common sources of germs. They can easily 4._____ other foods. After handling these foods, always wash your hands, utensils and surfaces thoroughly before you touch anything else.

Rule 2: One important way of stopping cross-infection is to make sure that you always 5._____ for your raw meat and cooked foods.

Rule 3: After use, wash all your dishes and utensils with 6._____. Change the water regularly, then 7._____. When possible, leave every thing to drain until dry.

Part C   Kitchen Safety

Learning Goals

You will be able to:
1. know the dangerous things in the kitchen;
2. establish safety awareness in the kitchen production.

Vocabulary Assistance

priority [praɪˈɒrəti]   n. 优先,优先权
hazard [ˈhæzəd]   n. 危害,危险,障碍
take precautions   采取预防措施
injury [ˈɪndʒəri]   n. 伤害,损害
accident [ˈæksɪdənt]   n. 事故,意外
maintain [meɪnˈteɪn]   v. 维持,继续
filter [ˈfɪltə(r)]   n. 滤网
vent [vent]   n. 进出口,通风口
duct [dʌkt]   n. 通风道,管道
grease [griːs]   n. 油脂
severe [sɪˈvɪə(r)]   adj. 严重的
rag [ræɡ]   n. 破布

oily [ˈɔɪli]   adj. 油的,油质的
soapy water   肥皂水
deep fat fryer   深油炸锅
foam [fəʊm]   n. 泡沫
bubble [ˈbʌbl]   n. 气泡
supervise [ˈsuːpəvaɪz]   v. 监督
fire extinguisher   灭火器
damage [ˈdæmɪdʒ]   v. 损害,损毁
falling down   跌倒
slip [slɪp]   v. 滑动,滑倒
trip [trɪp]   v. 绊倒
slippery [ˈslɪpəri]   adj. 滑的

burn [bɜːn] v. 烧伤	immediately [ɪˈmiːdiətli] adv. 立即，立刻
endanger [ɪnˈdeɪndʒə(r)] v. 危及，使遭到危险	issue [ˈɪʃuː] n. 问题
spill [spɪl] v. 溅出，洒出	promptly [ˈprɒmptli] adv. 迅速地
wipe up 擦干净，擦掉	resolve [rɪˈzɒlv] v. 解决
sprinkle [ˈsprɪŋkl] v. 撒，洒	uniform [ˈjuːnɪfɔːm] n. 制服
	blister [ˈblɪstə(r)] n. 水泡

Kitchen Safety

What is the most dangerous thing in the kitchen?

Safety is the number one priority in your kitchen.

When working or preparing food in the kitchen, one must be aware of safety hazards that may occur and take precautions in preventing injuries or accidents from happening by creating and maintaining a safe working environment.

First, clean all filters, vents and air conditioning ducts often. Why would people do that? Because of the oil or grease's high temperature, these types of burns can cause severe injury or death. Oily rags are dangerous. Wash them in soapy water or throw them away. Make sure all oil or grease burns receive prompt and proper medical treatment.

Using deep fat fryers is dangerous. Watch out for smoke, foam, bubbles, or a temperature above 370 °F. You need to be taught how to put out an oil or grease fire. Do not use water on an oil or grease fire. Do not use water on an electrical fire. You need to have carefully supervised practice in putting out oil and grease fires with a proper type of chemical fire extinguisher.

Damaged electrical wires are dangerous. Also do not use damaged electrical equipment.

In a busy kitchen, water or grease on the floor is quite dangerous. Falling down can be quite dangerous. You will slip and fall because of a slippery floor, and you can get badly hurt. You need to avoid slipping or tripping. If you are carrying something hot, you can also get badly burned. If you are carrying something made out of glass or something sharp, you can also get a deep cut. Don't spill the water on the ground. Pick up everything you drop. Wipe up everything you spill. Your kitchen floor must always be dry, otherwise you endanger yourself and your co-workers. Wipe up the oil as best as you can. Then sprinkle salt on the floor.

Last, report all accidents, safety issues and dangerous conditions immediately to your supervisor. Make sure these issues are promptly resolved.

Todd: Now, Rebecca, we're talking about working in the kitchen. I was a waiter and when I would help out in the kitchen, I was always afraid of the big knives and the fires, so can you talk a little about safety and maybe about some injuries you had working in the kitchen?

Rebecca: Yeah, that's actually really important. First there's the uniform. You have to cover as

外教有声

much of your skin as you can to avoid burns. If you have a special chef jacket, it must be all cotton so if you get something hot on it, it will still be safe, and it can be quickly taken off, so if you spill something very hot on your clothes, you actually remove the top layer and then you have something underneath, so you can avoid the hot thing being on your skin.

Todd: Well, have you ever been burned, and when you are burned what do you do to your skin to make the burn go away?

Rebecca: Yeah, I've only ever had one bad burn. It was from pork fat and I burnt my arm. I had a horrible blister afterwards, but you have to be careful not to touch the burn or break it. You should of course immediately put in under cold water, and then I use vitamin E and that was really good.

Todd: What about cuts? I imagine that you must have a million cuts from all those big sharp knives. What do you do for that?

Rebecca: Actually, I've never cut myself. No! Because they teach you when you learn how to chop a way to keep all your fingers out away from the knife, and you always have the knife in contact with your hands, so you don't need to look at it when you cut. You can feel where the knife is.

Todd: That's pretty impressive. Wow! OK, last thing. I guess the only danger I would see in the kitchen is just slipping and falling. The floor is always wet and greasy or whatever. What do you do about that?

Rebecca: You wear really, really heavy boots. I had a huge pair of boots, and of course we clean the floors really carefully. At the end of every shift, you get rid of as much grease as possible and we use non-slip mats, so that helps.

Todd: Cool. Thanks for the safety tips, Rebecca. Thanks.

Exercises

Activity I

Signs of Safety

Directions: Identify the practices described below as safe or dangerous. If the practice is safe, write "safe" in the space provided. If the practice is dangerous, write "danger", and explain why the practice is dangerous.

_____ 1. Jody put her long hair into a ponytail before she went into the kitchen to cook supper.

_____ 2. Debra chose to use knives that were not very sharp so that she would not cut herself.

_____ 3. The oven was so dirty that Elise decided to mix two strong cleansers in order to clean it.

_____ 4. Clare had the fire extinguisher removed because she did not like the way it looked in the kitchen.

_____ 5. Beverly stood to the side when she opened the oven door.

Activity II
Safety first, how much do you already know about kitchen safety? For each question below, decide whether the practice is safe or unsafe.

1. Use a towel or your apron to remove a pan from the oven.

2. Pour salt or baking soda over the flames of a grease fire.

3. Wipe up the spills on the floor right away.

4. Pour water on a grease fire.

5. Tie back long hair.

6. Climb up on the counter to get items from the top shelf.

7. Use electric appliances with wet hands.

8. Wear loose clothing while working in the kitchen.

9. Cut away from your body when using a sharp knife.

10. Keep cabinet doors open so everything is in easy reach.

safe：_____

unsafe：_____

Focus on Language

Kitchens can be 1._____ places. There are a lot of 2._____, such as knives. There are contaminants and bacteria that can be anywhere in the kitchen. Often the 3._____ in the kitchen is heavy and things can get slipperier if someone spills something on the floor and fails to clean it up.

Storing the food at the proper temperature and in proper 4._____ is essential to preventing food poisoning. Store all cold 5._____ in the refrigerator immediately after meals or purchase. Store all dry items in cabinets in air-tight plastic containers. Freeze all meat that are not going to be used in 6._____. Do not allow hot food to cool or cold food to warm before serving.

7._____ can be very dangerous. A spill on the floor can cause people to slip and get hurt. A spill on a counter can make utensils and bowls slippery which can cause injury if the object is dropped or comes into contact with the body. This is especially true for heavy bowls and knives.

Unit 3

Tools and Equipment

 Part A Hand Tools

 Learning Goals

You will be able to:
1. tell the name of hand tools in the kitchen;
2. know the use of hand tools in the kitchen.

 Vocabulary Assistance

ball cutter [bɔːl ˈkʌtə(r)]　挖球器
handle [ˈhændl]　n. 拉手,把手
　　　　　　　　v. 应付,应对
spatula [ˈspætʃələ]　n.（搅拌或涂敷用的）铲,抹刀,刮刀
palette knife [ˈpælət naɪf]　调色刀
blade [bleɪd]　n. 刀片
spread [spred]　v. 展开,伸开,传播,涂,分摊
　　　　　　　　n. 散布,广泛
rubber [ˈrʌbə(r)]　n. 橡胶,橡皮擦
　　　　　　　　adj. 橡胶做的
plastic [ˈplæstɪk]　n. 塑料
　　　　　　　　adj. 塑料的
scrape [skreɪp]　v. 刮,擦
stiff [stɪf]　adj. 严厉的,僵硬的
dough [dəʊ]　n. 生面团
pastry [ˈpeɪstri]　n. 糕点,油酥糕点,油酥面团,油酥面皮
wheel [wiːl]　n. 轮子,旋转

skimmer [ˈskɪmə(r)]　n. 撇取者,撇取物
stock [stɒk]　n. 股份,股票,库存,树干,家畜,高汤
colander [ˈkʌləndə(r)]　n. 滤锅,漏勺
French fries [frentʃ fraɪz]　n. 炸薯条
sieve [sɪv]　n. 筛子,滤网
　　　　　　v. 筛,筛选,滤
sift [sɪft]　v. 筛选,滤
grater [ˈɡreɪtə(r)]　n. 擦菜板
grate [ɡreɪt]　v. 磨碎,压碎
tongs [tɒŋz]　n. V 形夹子
whisk [wɪsk]　n. 搅拌器
　　　　　　v. 搅拌
tube [tjuːb]　n. 管子,管状物
rolling pin [ˈrəʊlɪŋ pɪn]　擀面杖
slicer [ˈslaɪsə]　n. 切片机
corkscrew [ˈkɔːkskruː]　n. 瓶塞钻,螺丝锥
meat hammer [ˈhæmə(r)]　肉锤
beat [biːt]　v. 击打

 Text

1 Ball cutter, melon ball scoop

The blade is a small, cup-shaped half-sphere. They are used for cutting fruits and vegetables into small balls.

2 Cook's fork

A heavy, two-pronged fork with a long handle used for lifting and turning meat and their items.

3 Straight spatula or palette knife

A long, flexible blade with a rounded end. It is mostly used for spreading icing on cakes and for mixing.

4 Rubber spatula

A broad, flexible rubber or plastic tip on a long handle used to scrape bowls and pans.

5 Bench scraper or dough knife

A broad, stiff piece of metal with a wooden handle on one edge. It is used to cut pieces of dough.

6 Pastry wheel or wheel knife, pizza cutter

A round, rotating blade on a handle used for cutting rolled-out dough, pastry and pizza.

7 Skimmer

It is used for skimming forth from liquid and for removing solid pieces from soup, stock and other liquid.

8 Colander

Bowl-shaped pan with many small holes in the bottom used for separating liquid from food.

9 Frying basket

The basket which used to fry the French fries. It can be used at high temperature.

10 Sieve

A tool of wire or plastic net on a frame used for separating large from small solid pieces, or solid things from liquid.

11 Grater

A kitchen utensil with a rough surface used for grating food.

12 Tongs

A tool used to pick up and handle food.

13 Whisk

A tool used to mix liquid, eggs, etc. into a stiff light mass.

14 Roasting fork

When roast food, the fork used for getting the food or sticking into the food to check whether the food is cooked.

15 Chopping board

A board made of wood or plastic used for cutting meat or vegetables.

16 Pastry bags and tubes

They are used for shaping and decorating with items such as cake icing, whipped cream and soft dough.

17 Pastry brush

The tool is used to brush items with egg wash, glaze, etc.

Can opener

It is used for opening food cans and tins.

 Mixing bowl

The round shape bowl used for mixing ingredients.

⑳ **Rolling pin**

A kitchen tool made of a long, thin and round piece of wood used for making dough flat.

㉑ **Slicer**

The tool is used for slicing different food.

㉒ **Corkscrew**

A bottle opener that pulls corks.

㉓ **Meat hammer**

A hammer used for beating meat.

Dialogue

Commis: Shall I beat eggs?

Chef: Of course! Use whisk to beat eggs in a mixing bowl.

Commis: And then?

Chef: Mix eggs, flour, melted butter, yeast with rubber spatula, we will make the dough. Remember flour should be sifted with sieve.

Commis: OK. What should I do next?

Chef: Cut the dough into pieces with bench scraper.

Commis: Are we going to make bread?

Chef: Yes.

Exercises

Activity Ⅰ What tools can you see in the pictures? Please write them down, then translate them into English.

1._____

2._____

3._____

4._____

5._____

6._____

Unit 3　Tools and Equipment

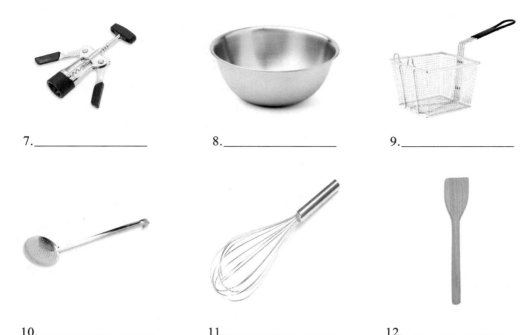

7.＿＿＿＿＿＿＿＿　　8.＿＿＿＿＿＿＿＿　　9.＿＿＿＿＿＿＿＿

10.＿＿＿＿＿＿＿＿　　11.＿＿＿＿＿＿＿＿　　12.＿＿＿＿＿＿＿＿

Activity Ⅱ　Match the phrases in column B with the kitchen tools in column A, and then make a sentence orally.

Example: I want to grate cheese with the grater.

A	B
rolling pin	cut meat or vegetables
meat hammer	cut pieces of dough
chopping board	beat meat
frying basket	make French fries
bench scraper	make dough flat

Learning Goals

You will be able to:

1. tell the name of the pots and pans;
2. know the use of the pots and pans.

stockpot ['stɒkpɒt]　n. 汤锅　　　　　　　sear [sɪə(r)]　v. 烧焦

brown [braʊn]　n. 棕色
　　　　　　　v. 使变成棕色
　　　　　　　adj. 棕色的
simmer ['sɪmə(r)]　n. 炖
　　　　　　　　v. 炖，煨
sauté ['səʊteɪ]　v. 炒
brazier ['breɪziə(r)]　n. 焖锅
braise [breɪz]　v. 炖，焖
poultry ['pəʊltri]　n. 家禽
roast [rəʊst]　v. 烤
　　　　　　n. 烤肉
wok [wɒk]　n. 炒菜锅
stir-fry ['stɜ:fraɪ]　v. 用旺火炒
　　　　　　　　n. 炒菜
double boiler ['dʌbl 'bɔɪlə(r)]　双层蒸锅
upper ['ʌpə(r)]　adj. 较高的
sit [sɪt]　v. 坐，放置
absorb [əb'sɔ:b]　v. 吸收
grease [gri:s]　n. 动物油脂，润滑油
　　　　　　v. 涂油脂于，用油脂润滑

crepe [kreɪp]　n. 薄煎饼
measure ['meʒə(r)]　v. 衡量，测量
batter ['bætə(r)]　n. 面糊
flip [flɪp]　v. 轻弹，轻击
　　　　　n. 跳跃，轻抛
slide [slaɪd]　v. 滑动
knob [nɒb]　n. 小块
dice [daɪs]　n. 骰子
　　　　　v. 将……切成丁
ribbon ['rɪbən]　n. 带，缎带，带状物
　　　　　　　v. 把……撕成条带
handful ['hændfʊl]　n. 少数，少量，一把（的量）
season ['si:zn]　n. 季节
　　　　　　　v. 调味
squeeze [skwi:z]　v. 挤压，压榨
smoked salmon [sməʊkt 'sæmən]　烟熏三文鱼
chowder ['tʃaʊdə(r)]　n. 杂烩

Text

❶ Frying pan

It is a flat-bottomed pan used for frying, searing and browning food.

❷ Stockpot

A large, deep and straight-sided pot used for preparing stocks and simmering large quantities of liquid.

❸ Sauce pan

A small deep cooking pan with a handle used for boiling, stewing and making sauces.

❹ Sauté pan

It is used for general sautéing and frying meat, fish, vegetables and eggs.

❺ Brazier

A round, broad, shallow and heavy-duty pot with straight sides used for browning, braising and stewing meat.

❻ Roasting pan

It is used for roasting meat and poultry.

❼ Wok

A round-bottomed steel pan with two loop handles used for stir-fry, especially Chinese cuisine.

❽ Double boiler

A pot with two sections. The lower section holds boiling water, and the upper section holds food.

 **Exercises**

Activity I Read this following recipes and translate them into Chinese.

Crepe Recipe

Ingredients:

4 eggs

1 cup of flour

1/2 cup of milk

1/2 cup of water

1/2 teaspoon of salt

2 tablespoons of melted butter

Preparation:

1. Measure all ingredients into blender jar, blend for 30 seconds.

2. Scrape down sides. Blend for 15 seconds more. Cover and let sit for 1 hour(this helps the flour absorb more of the liquid).

Regular Crepe Pan Directions:

1. Heat crepe pan and grease lightly.

2. Measure about 1/4 cup of batter into pan. Tilt pan to spread batter. Once crepe has lots of little bubbles, loosen any edges with the spatula. Flip crepe over. This side cooks quickly. Slide crepe from pan to plate.

Smoked Salmon Chowder

Ingredients:

a knob of butter

1 onion, finely diced

750 grams of potatoes, diced

500 milliliter of chicken stock

500 milliliters of milk

350 grams of smoked salmon, cut into ribbons

a handful of parsley leaves, finely chopped

1 lemon, halved

Methods:

1. Fry the onion gently with butter in the sauté pan, then add potatoes, stock and milk, and simmer until potatoes are very tender.

2. Add the smoked salmon and parsley leaves, and season well.

3. Heat everything through and add a squeeze of lemon.

Activity II What pans can you see in the pictures? Please write them down, then translate them into English.

1. _____

2. _____

3. _____

4. _____

5. _____

Part C Knives

Learning Goals

You will be able to:
1. tell the name of different knives;
2. know the use of different knives.

Vocabulary Assistance

chef's knife 厨师刀
frequently ['fri:kwəntli] adv. 频繁地，经常
purpose ['pɜːpəs] n. 意图，计划
chop [tʃɒp] v. 砍，切
　　　　　　 n. 排骨，劈
paring knife ['peərɪŋ naɪf] 削皮刀，去皮刀
pointed ['pɔɪntɪd] adj. 尖的，尖锐的，严厉的，直截了当的，突出的
blade [bleɪd] n. 刀片，叶片，桨叶
inch [ɪntʃ] n. 英寸
raw [rɔː] adj. 生的，自然状态的

boning knife ['bəʊnɪŋ naɪf] 拆骨刀，剔肉刀
serrated knife [se'reɪtɪd naɪf] 锯齿刀
butcher knife ['bʊtʃə(r) naɪf] 屠刀
peeler ['piːlə(r)] n. 去皮器，削皮器
slotted ['slɒtɪd] adj. 开槽的
swiveling ['swɪvlɪŋ] adj. 转动的
cleaver ['kliːvə(r)] n. 砍肉刀，剁肉刀
steel [stiːl] n. 钢，钢铁，钢制品
maintain [meɪn'teɪn] v. 维持，保持，维修，保养

Text

❶ French knife or chef's knife

It is the most frequently used knife in the kitchen used for general-purpose chopping, slicing, dicing, and so on.

❷ Paring knife

A tool with a small, 2-4 inches long pointed blade used for trimming and paring vegetables and fruits.

❸ Boning knife

A tool with a thin, about 6 inches long pointed blade used for boning raw meat and poultry.

❹ Serrated knife

It is like a slicer, but with a serrated edge used for cutting bread, cakes and similar items.

❺ Butter knife

It is used for spreading butter and jam.

❻ Butcher knife

A tool with a heavy, broad and slightly curved blade used for cutting and sectioning raw meat.

❼ Vegetable peeler

A short tool with a slotted and swiveling blade used for peeling vegetables and fruits.

❽ Steel

It is an essential part of the knife used for maintaining knife edges.

❾ Chinese cook's knife

It is used for Chinese cuisine.

❿ Cleaver

A tool with a heavy and broad blade used for cutting through bones.

⓫ Carving knife

It is used for carving food.

Exercises

Activity What knives can you see in the pictures? Please write them down, then translate them into English.

1._____   2._____   3._____

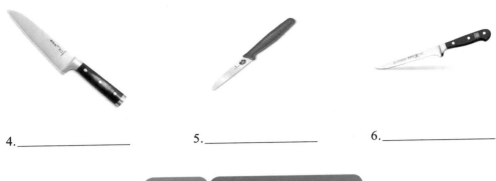

4. _____ 5. _____ 6. _____

Part D Cooking Equipment

Learning Goals

You will be able to:

1. tell the name of different cooking equipment;
2. know the use of different cooking equipment.

Vocabulary Assistance

deep fryer [di:p 'fraɪə(r)] 油炸锅
process ['prəʊses] n. 过程
　　　　　　　　　v. 加工处理
versatile ['vɜːsətaɪl] adj. 多用途的,多才多艺的
valuable ['væljuəbl] adj. 有价值的,宝贵的
evenly ['iːvnli] adv. 均匀地
uniformly ['juːnɪfɔːmli] adv. 一致地,相同地
mixer ['mɪksə(r)] n. 搅拌器
blender ['blendə(r)] n. 搅拌器
food processor [fuːd 'prəʊsesə(r)] 食品加工机
electric [ɪ'lektrɪk] adj. 用电的,电的
interchangeable [ˌɪntə'tʃeɪndʒəbl] adj. 可互换的
blade [bleɪd] n. 刀片
container [kən'teɪnə(r)] n. 容器
toaster ['təʊstə(r)] n. 烤面包机
fragrant ['freɪɡrənt] adj. 芳香的,香的
egg beater [eg 'biːtə(r)] 打蛋器

egg white 蛋清
egg yolk [jəʊk] 蛋黄
foam [fəʊm] n. 泡沫
　　　　　　 v. 起泡沫
ice maker [aɪs 'meɪkə(r)] 制冰机
multifunctional [ˌmʌlti'fʌŋkʃənl] adj. 多功能的
steam [stiːm] n. 水蒸气
braise [breɪz] v. 炖,焖
iron ['aɪən] n. 熨斗,铁制品
　　　　　　 v. 熨烫
　　　　　　 adj. 铁质的
automatic [ˌɔːtə'mætɪk] adj. 自动的
steak [steɪk] n. 牛排,肉排
knead [niːd] v. 捏,揉
pasta ['pæstə] n. 意大利面食
fermenting box [fər'mentɪŋ bɒks] 醒发柜
dough [dəʊ] n. 生面团
obtain [əb'teɪn] v. 得到,存在
edible ['edəbl] adj. 可食用的

Text

❶ Deep fryer

A deep fryer has only one use: to cook food in hot fat.

❷ Slicer

The slicer is a valuable machine because it slices food more evenly and uniformly than can be done by hand.

❸ Mixer

It is like a large, powerful and high-speed blender. It is used to chop and mix large quantities of food rapidly.

4 **Food processor**

It is an electric kitchen appliance with a set of interchangeable blades revolving inside a container.

5 **Toaster**

It is a special electric cooker for re-baking sliced bread. It can not only bake bread slices into yellow, but also make them more fragrant and taste better.

6 **Egg beater**

A tool used to mix egg whites and egg yolks into egg liquid and separate egg whites and egg yolks to foam.

7 **Ice maker**

It is used to make ice for making food and drink.

⑧ Oven
It is used for baking and roasting.

⑨ Multifunctional steam oven
It is used for baking, steaming, frying, boiling and braising.

⑩ Western style oven
The oven has a high degree of automatic control, easy to operate, and is used for baking dishes.

⑪ Western style grill

It is used to make steak, pork chop, lamb chop, fried eggs, etc.

⑫ Pressure machine

It is a food machine that mixes flour with water and replaces traditional hand kneading. It can be used to make noodles, cakes, pasta, and so on.

⑬ Fermenting box

It is used to make the dough produce gas and expand again, so as to obtain the volume needed to make the steamed bread and the finished bread have better edible quality.

Exercises

Activity I   Read this following recipe and translate them into Chinese.

Lady Fingers

Ingredients:

4 eggs, separated

2/3 cup of white sugar

7/8 cup of all-purpose flour

1/2 teaspoon of baking powder

Methods:

1. Preheat oven to 200 ℃. Prepare two 17 inches × 12 inches baking sheets with baking parchment. Fit large pastry bag with a plain 1/2 inch round tube.

2. Place egg whites in the bowl and beat on high speed. Slowly add 2 tablespoons of sugar and continue beating until stiff and glossy. In another bowl beat egg yolks and remaining sugar. Whip until thick and very pale in color.

3. Sift flour and baking powder together on a sheet of wax paper. Fold half the egg whites into the egg yolks mixture. Fold in flour, and then add the remaining egg whites. Transfer mixture to the pastry bag and pipe out onto prepared baking sheet. Bake 8 minutes.

Activity II Match the phrases in column B with the kitchen equipment in column A, and then make a sentence orally.

Example: I want to grate cheese with the grater.

A	B
whisk eggs	deep frier
French fries	food processor
make the soup muddy	pressure machine
flatten the dough	oven
bake bread and cakes	egg beater

Unit 4

Condiments and Spices

Part A Condiments

 Learning Goals

You will be able to:
1. get familiar with condiments;
2. know the English names of condiments;
3. know how to classify condiments.

 Vocabulary Assistance

soy sauce　酱油
mustard ['mʌstəd]　n. 芥末
vinegar ['vɪnɪɡə(r)]　n. 醋
ketchup ['ketʃəp]　n. 番茄酱
maple syrup　枫糖浆
soybean paste　黄酱
curry ['kʌri]　n. 咖喱
rock sugar　冰糖
wasabi　青芥末
horse-radish　辣根
vinaigrette [ˌvɪnɪ'ɡret]　n. 油醋汁
oyster oil　蚝油
Tabasco [tə'bæskəʊ]　n.（塔巴斯科）辣椒酱
chili paste　辣酱
do the trick　达到理想的结果，做成功
refine [rɪ'faɪn]　v. 精炼，提纯
purify ['pjʊərɪfaɪ]　v. 净化，使纯净
concentrate ['kɒnsntreɪt]　v. 集中，浓缩
nutrient ['njuːtriənt]　n. 营养素
vitamin ['vɪtəmɪn]　n. 维生素，维他命
mineral ['mɪnərəl]　n. 矿物质
fiber ['faɪbə(r)]　n. 纤维，纤维素

amino acid [ə'miːnəʊ 'æsɪd]　氨基酸
molasses [mə'læsɪz]　n. 糖蜜，糖浆
dark brown sugar　红糖，黑糖
light brown sugar　黄砂糖，黄糖
cornstarch ['kɔːnstɑːtʃ]　n. 玉米淀粉
clump [klʌmp]　v. 使成一丛，使凝结成块
demarara sugar　金砂糖
muscovado [ˌmʌskə'vɑːdəʊ]　n. 黑砂糖，粗糖
marinade [ˌmærɪ'neɪd]　n. 腌泡汁
demerara　（产于加勒比地区的）一种红糖
comb [kəʊm]　n. 蜂巢
soothe [suːð]　v. 安慰，安抚
scratchy ['skrætʃi]　adj. 刺痛的，发痒的
antibacterial [ˌæntibæk'tɪəriəl]　adj. 抗菌的
antiseptic [ˌænti'septɪk]　adj. 防腐的
ulcer ['ʌlsə(r)]　n. 溃疡
authentic [ɔː'θentɪk]　adj. 真正的，真实的
commercial [kə'mɜːʃl]　adj. 商业的
glaze [ɡleɪz]　v. 变得光滑
saline ['seɪlaɪn]　adj. 含盐的，咸的
acid ['æsɪd]　adj. 酸的
spicy ['spaɪsi]　adj. 辛辣的，加有香料的

Sugar and Syrup

Next time you go for coffee or tea at a cafe, check out the sugar options. There may be white sugar, brown sugar, sugar "in the raw", honey and perhaps even sugar syrup. When it comes to sweetening your coffee or tea, they'll all do the trick.

But when it comes to cooking and baking, you really need the right one for the job. Even if you're just making chocolate chip cookies, using brown sugar (rather than white sugar) will help turn out a tastier and chewier cookie.

The sugar we know best is made from sugar cane and sugar beets. The refining process removes the natural sugar stored in the raw plant material. The sugar is purified, filtered, concentrated and dried to produce the sweet stuff we put in coffee or in baked goods. The refining process removes the natural color of sugar, and all of the nutrients. White sugar has "empty calories" which do give you an energy rush, but it lacks nutrients such as vitamins, minerals, amino acids as well as fiber.

Brown sugar is white sugar that has been mixed with molasses. It's available in both light and dark versions, the dark containing more molasses. Recipes will sometimes specify light or dark brown sugar, but in general, you can use whichever you prefer.

Icing sugar is white sugar that has been ground into powder and mixed with a bit of cornstarch to keep it from clumping. A sprinkle of this snow-white powder puts a beautiful finishing touch on French toast and many other treats.

These days, the sugar is not actually "raw", because it undergoes a process to remove contaminants. There are various other kinds of raw sugar:

Demarara comes from Guyana, has a coarse texture and toffee flavor.

Muscovado or Barbados sugar is finer than demerara, with a strong molasses flavor. It's also added to coffee or tea, and used in fruit cakes, marinades and sauces. Substitute with dark brown sugar.

Honey is sweeter than sugar, so you can use less of it for the same result. Liquid honey is what's used in baking, although you can also buy honey in the comb. Mixed with hot water and lemon, honey can help soothe a scratchy throat. Honey has traditionally been used for antibacterial and antiseptic purposes, like treating topical wounds and stomach ulcers.

Maple syrup comes from the sap of maple trees. There's a big difference in taste between authentic maple syrup and commercial pancake syrup. Maple syrup is often used as a glaze for salmon, in marinades and as an ice cream flavoring.

Well that's about enough sweet-talking for one day!

外教有声

Lisa: Do you like cooking, Julia?

Julia: I really enjoy it, especially when it ends up tasting good!

Lisa: What kind of dishes do you usually make?

Julia: I almost always make either a beef roast or a chicken roast with asparagus, parsnips, peas, carrots and potatoes on Sunday.

Lisa: How about spicy food?

Julia: My family loves spicy food. We often eat Chinese, Thai, Indian, or Mexican food when were in the mood for spice.

Lisa: What's your favorite dish to make?

Julia: I absolutely love making Mapo Tofu, which is a Chinese dish with bean curd. In order to cook Chinese delicious dishes, I bought a lot of condiments.

Lisa: Oh, what do you have?

Julia: I have soy sauce, oyster oil, soybean oil, chili paste, vinegar, etc.

Exercises

Activity Ⅰ   Please place the right words from the box and write down in the blanks.

soy sauce	mustard	vinegar	ketchup	chili paste	maple syrup
fish sauce	curry	rock sugar	wasabi	horse-radish	vinaigrette
oyster oil	honey	Tabasco			

1._____

2._____

3._____

4._____

5._____

6._____

7._____

8._____

9._____

Unit 4　Condiments and Spices

10._____　　11._____　　12._____

13._____　　14._____　　15._____

Activity Ⅱ　Can you identify different kinds of condiments? Put the right words into the box from above pictures.

Variety	Condiments
Saline seasonings	
Acid seasonings	
Hot seasonings	
Sweet seasonings	

Focus on Language

 Read the following passage and answer the questions.

Soy Sauce

A condiment originally from China, soy sauce occupies a preeminent position in the cuisines of Asian countries. Its Japanese name is shoyu.

Traditionally, soy sauce, shoyu and tamari refer to the liquid formed during the manufacture of miso.

Traditional Chinese soy sauce is made from whole soybeans and ground wheat. It can be more or less dark depending on its age and whether caramel or molasses has been added. Tamari is made exclusively from soybeans or soybean meal (the residue from pressing the beans when oil is extracted); therefore, it contains no cereal grain. It sometimes contains additives such as monosodium glutamate and caramel. Tamari is dark and has a thicker consistency. Shoyu is lighter in color than Chinese soy sauce and slightly sweet.

Soy sauce (Chinese or Japanese) contains some of the alcohol produced during the fermentation

外教有声

of the cereal grains, whereas tamari has none. The soy sauce found in the supermarket is usually a synthetic product that is a pale imitation of the original.

Question 1: Which part of the world is soy sauce often used?
Question 2: What is the difference between Chinese soy sauce and Japanese shoyu?

Part B Spices

Learning Goals

You will be able to:
1. know the English names of spices;
2. know how to classify spices;
3. know the characteristics of spices.

Vocabulary Assistance

pepper ['pepə(r)]   n. 胡椒
spice [spaɪs]   n. 香料
herb [hɜːb]   n. 香草
enhance [ɪn'hɑːns]   v. 提高,增强
retain [rɪ'teɪn]   v. 保持
accommodation [əˌkɒmə'deɪʃn]   n. 住处,膳宿
wrap [ræp]   v. 包起来
basil ['bæzl]   n. 罗勒
chili pepper   红辣椒
bay leaf   月桂叶,香叶
caraway seed   页蒿子
chervil ['tʃɜːvɪl]   n. 细叶芹
chives [tʃaɪvz]   n. 细香葱
cilantro [sɪ'læntrəʊ]   n. 香菜叶
dill [dɪl]   n. 莳萝
fennel ['fenl]   n. 茴香
savory ['seɪvərɪ]   n. 香薄荷
legume ['leɡjuːm]   n. 豆类

cinnamon ['sɪnəmən]   n. 肉桂
rosemary ['rəʊzmərɪ]   n. 迷迭香
nutmeg ['nʌtmeɡ]   n. 肉豆蔻
paprika [pə'priːkə]   n. 甜椒粉
cumin ['kʌmɪn]   n. 小茴香
turmeric ['tɜːmərɪk]   n. 姜黄,姜黄根粉
thyme [taɪm]   n. 百里香
sesame seed   芝麻,芝麻籽
clove [kləʊv]   n. 丁香
Sichuan pepper   花椒
mint [mɪnt]   n. 薄荷
star anise   八角茴香
sage [seɪdʒ]   n. 鼠尾草
parsley ['pɑːslɪ]   n. 荷兰芹,欧芹
marjoram ['mɑːdʒərəm]   n. 马郁兰
oregano [ˌɒrɪ'ɡɑːnəʊ]   n. 牛至
tarragon ['tærəɡən]   n. 龙蒿
saffron ['sæfrən]   n. 藏红花

Using Herbs and Spices

Herbs and spices can be used to enhance and balance the flavor of your dishes, and make something taste more delicious that would otherwise be quite plain.

However, it's important to achieve the correct amount of seasoning so that your meal is not completely overpowered.

If they are fresh, herbs should generally be added to food toward the end of its cooking time, or sprinkled on top of the finished dish, so the flavor don't end up being cooked out and become tasteless.

Dried herbs should be added earlier in the process. Dried herbs will keep for a long time if you keep them out of direct light and away from any heat sources, and keep them in a cool place below room temperature.

If your accommodation has a garden or you have a large windowsill that gets lots of light, you can grow your own herbs, though they don't last as long as dried herbs, and are not as strong in flavor. This is why you will always need larger quantities if you cook with fresh herbs rather than dried herbs.

Sometimes your recipe will state to use fresh herbs and you will only have dried ones in stock. The general rule here is to use only a third as much of a dried herb than the amount of fresh herb in the recipe.

If you decide to buy fresh herbs or grow your own, they should be stored by wrapping them loosely in damp paper. You can also help retain their freshness by then placing them in plastic bags.

To give you an idea of what herbs and spices are available to you in cooking, we've put together the following list of some of the more common ones and the food they can be used to enhance.

Basil: Italian dishes, especially tomatoes and tomato sauce; pasta, chicken, fish, shellfish, soup.
Bay leaf: stews, tomato dishes, soup, sauce.
Chili pepper: soup, rice, sauce, meatloaf, bean or meat stews.
Caraway seed: cooked vegetables, e. g. potatoes, turnips, cabbage and carrots.
Chervil: fish, shellfish, chicken, turkey, mixed green salad, French dishes, vegetable dishes.
Chives: omelet, pasta, beef, turkey, chicken, soup, sauce, mixed green salad, seafood.
Cilantro: salad, chicken, turkey, rice, beans, shellfish, fish, Asian dishes.
Dill: chicken, lamb, salmon, vegetables, sauce, dressings.
Fennel: fish, curry, soup, beef, chicken, pasta, vegetable dishes.
Marjoram: tomato dishes, soup, stews, eggs, beef, chicken, turkey, fish, beans.
Nutmeg: fruit dishes, stewed beef and poultry, custard, eggnog, baked food and sweet bread.
Oregano: beef, lamb, fish, pasta, sauce, soup, stews, Italian and Greek dishes.
Paprika: baked fish, goulash, soup, stew, beans and legumes.
Parsley: meat, poultry, fish, vegetables, stuffing, omelets, bread.
Rosemary: pasta, stuffing, vegetables, grilled beef and chicken, casserole, fish, salad.
Sage: soup, vegetable dishes, stuffing, rice, chicken, duck, pork.
Savory: beans, soup, grilled chicken.
Tarragon: fish, poultry, sauce, green vegetables, veal, shellfish, dressings.
Thyme: beans and legumes, vegetable dishes, stews, soup, tomato dishes, chicken, turkey, fish.

There are a lot of guidelines around to which seasonings go best with certain food, but it's also down to the individual what tastes good, so don't be afraid to try different herbs and spices with a particular dish.

Experiment is all part of the learning process, and eventually you will discover which seasonings you prefer the most with your meals.

外教有声

Dialogue

Henry: We need to make production, could you prepare something for me?

Eric: Yes, chef. What do you need?

Henry: First, pick up all the spices from the dry store, and then scale ingredients for this recipe. For two times.

Eric: OK, chef. I got bay leaves, cumin seeds, cinnamon sticks and nutmeg. But the clove is out of store.

Henry: All right. That's fine. Steamed rice is in the oven now. When it's done, mix with saffron.

Eric: How about the marinated chicken?

Henry: Check the taste. Make sure the turmeric flavour is not so strong.

Eric: It's good. But I think the pepper is not enough.

Henry: Let me check. Hum, add a little bit. Mix it well and bake it.

Eric: Time and temperature, chef?

Henry: For 20 minutes, 190 ℃, I will check when the time is up.

Eric: Yes, chef.

Exercises

Activity Ⅰ Please place the right words from the box and write down in the blanks.

rosemary	caraway seed	nutmeg	paprika	cumin	turmeric
thyme	sesame seed	clove	peppercorn	Sichuan pepper	fennel
mint	star anise	sage	parsley	marjoram	oregano
tarragon	dill	bay leaf	basil	saffron	cinnamon

1._____   2._____   3._____

4._____   5._____   6._____

Unit 4 Condiments and Spices

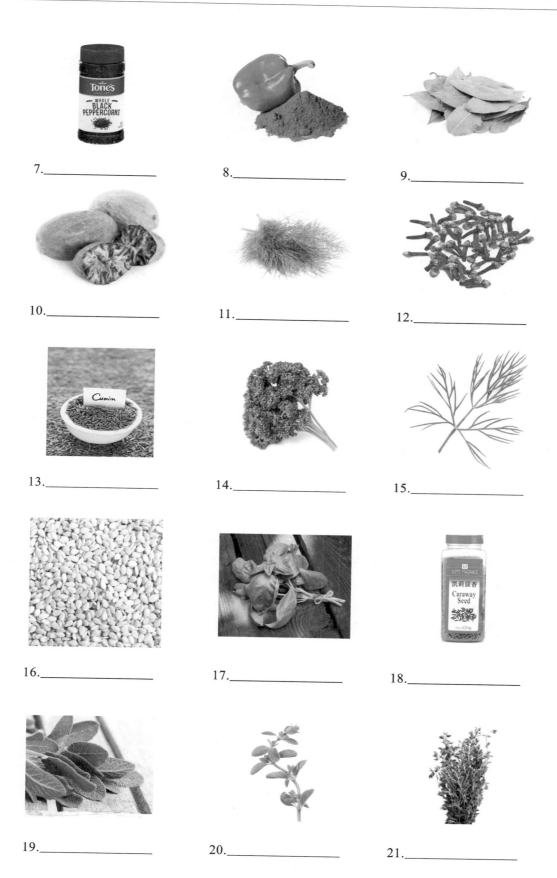

7. _____ 8. _____ 9. _____

10. _____ 11. _____ 12. _____

13. _____ 14. _____ 15. _____

16. _____ 17. _____ 18. _____

19. _____ 20. _____ 21. _____

22._____ 23._____ 24._____

Activity Ⅱ　Translate the following sentences into English.
1. 鱼露是沿海地区的传统发酵调味品。
2. 然后你把佐料放进面条里，再加点盐。
3. 日本芥末通常配生鱼片。
4. 这肉里应该用盐和芥末调味。
5. 在法国，芥菜籽被浸透，然后再磨成糨糊状。
6. 酸辣酱可以混合在任何一种印度餐中，形成一种不同的口味。
7. 我想在上面放一些火腿、香肠、蘑菇、洋葱、橄榄和菠萝。
8. 盐是一种常用的食物防腐剂。
9. 肉豆蔻常用作食物中的香料。
10. 这个汤的特殊香味是因为藏红花粉。

Focus on Language

🎧 **Listen to the tape and fill in the blanks with the missing words.**

There are about 35 spices which can be broadly classified into 1._____, based upon the parts of the plants which they are obtained, namely rhizomes and 2._____ and 3._____.

　　The basic classification of spices is as follows: First, leaves or branches of 4._____ plants, such as 5._____, tarragon, 6._____, oregano and chervil. Second, ripened fruits or seeds of plants, such as 7._____, fennel, coriander, 8._____, etc. Third, roots or bulbs of certain plants, such as 9._____ and 10._____.

Unit 5

Meat

Part A Understanding Meat of Domestic Animals and Wild Animals

扫码看课件

Learning Goals

You will be able to:
1. know the composition and classification of meat;
2. learn basic cutting methods and cooking methods of meat;
3. memorize the words or phrases of different body parts of domestic animals.

Vocabulary Assistance

domestic [dəˈmestɪk]　adj. 本国的,国内的,家用的,家庭的,家务的
quail [kweɪl]　n. 鹌鹑,鹌鹑肉
protein [ˈprəʊtiːn]　n. 蛋白质
carbohydrate [ˌkɑːbəʊˈhaɪdreɪt]　n. 碳水化合物
connective tissue　结缔组织
carcass [ˈkɑːkəs]　n. 动物尸体,(尤指供食用的)畜体
slice [slaɪs]　n. (切下的食物)薄片,片
　　　　　　v. 把……切成(薄)片,切
dice [daɪs]　n. (肉、菜等)丁,小方块
　　　　　v. 将(肉、菜等)切成小方块,将……切成丁
shred [ʃred]　v. 将(肉、菜等)切成丝,切碎
　　　　　　n. (撕或切的)细条,碎片
rib [rɪb]　n. 肋骨,排骨
sirloin [ˈsɜːlɔɪn]　n. 牛里脊肉,牛上腰肉
tenderloin [ˈtendəlɔɪn]　n. (牛、猪、羊的)里脊肉,嫩腰肉
plate [pleɪt]　n. (烹饪)侧胸腹肉
cattle [ˈkætl]　n. 牛
poultry [ˈpəʊltri]　n. 家禽,禽类的肉

mince [mɪns]　v. 用绞肉机绞(食物,尤指肉)
　　　　　　n. 绞碎的肉,肉末(尤指牛肉)
strip [strɪp]　v. (烹饪)把……切成条状
　　　　　n. (纸、金属、织物等)条,带
barbecue(BBQ) [ˈbɑːbɪkjuː]
　　　　　n. (户外烧烤用的)烤架,户外烧烤
　　　　　v. (在烤架上)烤,烧烤
broil [brɔɪl]　v. 烤,焙(肉或鱼),(使)变得灼热
sauté [ˈsəʊteɪ]　v. 炒,煸,嫩煎
fry [fraɪ]　v. 油炸,油煎
braise [breɪz]　v. 煨,炖
scald [skɔːld]　v. (不加任何调料地煮熟)白灼
chuck [tʃʌk]　n. 牛肩胛肉,上脑
shank [ʃæŋk]　n. (烹饪)牛腱,牛腿肉
beef round　牛(臀、腿)肉
beef brisket　牛胸肉
beef flank　牛肋腹肉
beef rump　牛后腿部的肉

Text A

As we talk about meat, not only do we refer to the flesh of domestic animals (cattle, pigs and lambs) and of wild animals, but also mean the flesh of poultry (geese, turkey, chicken, quails and ducks). Their composition and structure are essentially the same. Meat is muscle tissue which is composed of water (about 75 percent), protein (about 20 percent), fat (up to 5 percent) and other elements including carbohydrate in small quantities. Remember that meat consists of muscle fibers and connective tissue. Among them, protein is an important nutrient and the most abundant solid material in meat. Considering juiciness, tenderness and flavor, fat is necessary in the meat though the portion is not so high, of course, a beef carcass can be as much as 30 percent fat.

In order to make people enjoy delicious food better, make the food be good both in flavor and smell and in color and appearance, chefs pay great attention to cutting skills and cooking technologies. Raw meat can be cut into slices, segments, dices, chunks, shreds, cubes, grains, strips, and so on. Dry-heat methods and moist-heat methods are often used to apply heat to food in order to change them in ways we can control. In details, we often cook meat, poultry and even fish by roasting, baking, barbecuing, broiling, grilling, sautéing, pan-frying, deep-frying, stir-frying, boiling, simmering, shallow-poaching, steaming, smoking, braising, scalding, and so on.

When we enter a western-style restaurant, we can see fillet or tenderloin steak, rib-eye steak, sirloin or T-bone steak on the menu. In facts, those steaks are named according to different body parts of cattle. Let us see the picture below.

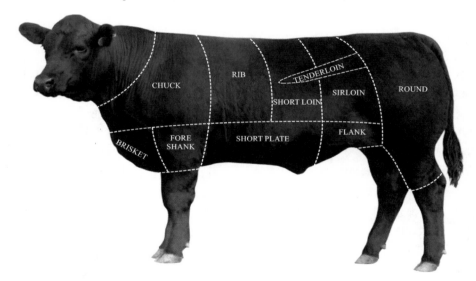

Dialogue

A: Chef, this beef is very good. How are you doing it?

B: By braising. It is quite easy.

A: Are there any other ways to cook the beef?

B: Certainly, beef can be cooked in different ways, like roasting, broiling, stewing, frying, and so on.

A: I see. But I want to know how we can apply these different ways in course of cooking?

B: It really depends. First of all, you must be familiar with the body parts. As you know, I cook this thick flank by braising. As for fillet and rib, you can roast them. Chuck is often stewed in our restaurants.

A: Thank you, chef. The cooking methods vary depending on the types of beef.

B: You bet. For the same type of meat, chefs can use several methods. For example, rump can be roasted and also be fried.

A: So it is with lamb and pork, right?

B: Of course, it is. You will be a top chef with more practice.

A: It is so kind of you, chef.

How to make "Sweet and Sour Pork Tenderloin"

Name of the dish: Sweet and Sour Pork Tenderloin

Ingredients: 200 grams of pork-tenderloin

Seasonings: 3 grams of scallions 3 grams of ginger
3 grams of minced garlic 100 grams of sugar
50 grams of rice vinegar 10 grams of soy sauce
1000 grams of edible oil 150 grams of starch
10 grams of soybeans

Cooking procedures:

1. Slice the pork tenderloin and put it in a bowl, and add some starch and stir well.

2. Add sugar and starch into another bowl, pouring vinegar, soy sauce, water and stir well.

3. Heat the edible oil at a temperature of 180 ℃ and deep-fry the sliced pork tenderloin up to medium well-done, make the pork brown and then drain it.

4. Pour little edible oil in the wok and add minced garlic, ginger and scallions, stir-fry them for a well and then pour the sauce, deep-fried pork tenderloin and green beans, stir-fry quickly, and then dish up.

Activity Ⅰ Look at the materials and dishes of beef/pork/lamb and translate them into Chinese.

1. Sautéed mutton with scallion _____
2. Beef fillet with black pepper _____

3. Twice-cooked pork _____

4. Honey-sauce meat _____

5. Stewed meat-ball _____

6. Sweet and sour pork tenderloin _____

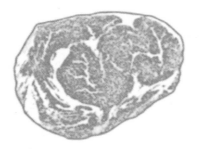
7. Rib eye steak _____

8. T-bone steak _____

9. Tenderloin steak _____

10. Flank steak _____

11. Heart _____

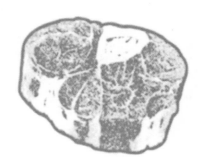

12. Shank _____

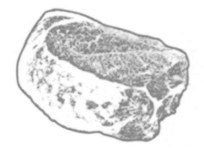

13. Boneless chuck _____

14. Sirloin steak _____

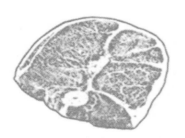

15. Round steak _____

16. Boneless rump _____

17. Brisket _____

18. Chuck _____

Activity Ⅱ Translate the following Chinese into English.

菜名:回锅肉

一、主辅料

猪连皮坐臀肉 200 克。

二、调味料

蒜苗 100 克,豆瓣酱 30 克,甜面酱 9 克,精盐 1 克,白糖 3 克,味精 1 克,酱油 3 克,色拉油 50 克。

三、制作工艺

1. 将猪肉放入凉水锅中用中火煮至断生后捞起晾凉(也可入冰箱冻 3~5 分钟,方便切片)。

2. 猪肉切成 6 厘米×4 厘米×0.2 厘米的大片,蒜苗切成马耳朵形。

3. 炒锅内放入色拉油,加热至 150 ℃,放入肉片炒至肉呈灯盏窝状,吐油时加精盐、郫县豆瓣酱炒香炒红,再加入甜面酱、酱油、白糖炒出香味,放入蒜苗炒至断生,最后放入精盐、味精炒匀,起锅装盘即成。

Focus on Language

 Listen to the following passage and fill in the blanks.

When you step into a western-style restaurant and 1._____ steak, the waiter will ask "How do you like your 2._____ cooked?" What does he mean? In fact, the 3._____ of cooked steak is influenced by how much it is done. Depending on the 4._____ for which the steak is cooked, it may be raw, 5._____, medium rare, medium, medium well-done and well-done. Rare steak is exposed to the 6._____ for a very short time. They still maintain their rawness and are very 7._____ in color. Rarely done steak maintains their 8._____ beefy flavor, but it is not very healthy as it still contains microorganisms. As the cooking time increases, the pinkness of steak gets converted to brownness and its juiciness also 9._____. Medium steak is pink in the center, grayish brown at the surroundings. Well-done steak is brown throughout and also 10._____ to chew.

Translation

Translate the following dialogue into Chinese.

(W=Waiter   C=Chef)

C: What did the customers order?

W: They wanted a rib-eye steak and French onion soup.

C: What did they want to accompany the steak?

W: With lettuce, onion and spaghetti.

C: Did they want it well-done, medium or rare?

W: Medium well-done. How long does it take?

C: It takes about 3 minutes.

W: OK, chef. I will inform them.

Part B Poultry

 Learning Goals

You will be able to:
1. know the process of making poultry;
2. tell materials and dishes of poultry;
3. memorize the words and phrases of different kinds of poultry.

 Vocabulary Assistance

versatility　n. 用途广泛,技术全面
diet-conscious ['daɪət'kɒnʃəs]　adj. 节食的
cholesterol [kə'lestərɒl]　n. 胆固醇
free-range　adj. 散养的
turkey ['tɜːki]　n. 火鸡
squab [skwɒb]　n. 乳鸽
farm-raised　adj. 农场养殖的
breast [brest]　n.（鸡、火鸡的）胸脯肉
on the contrary　恰恰相反,正相反,相反地
drumstick ['drʌmstɪk]　n. 鼓槌,熟鸡（或家禽）腿下段,下段鸡（或家禽）腿肉
thigh [θaɪ]　n. 大腿,股,食用的鸡（等的）大腿
stuffing　n. 填充物,填料
gut [gʌt]　n.（尤指动物的）内脏,（尤指大的）胃,肚子
　　v. 取出……的内脏（以便烹饪）
starch [stɑːtʃ]　n. 淀粉,含淀粉的食物

stock [stɒk]　n. 高汤,原汤
rack [ræk]　n. 支架,烧烤架
broiler ['brɔɪlə(r)]　n. 烤箱
drain [dreɪn]　v. 捞出,沥水
nutmeg ['nʌtmeg]　n. 肉豆蔻
fennel ['fenl]　n. 茴香
anise ['ænɪs]　n. 茴芹
clove [kləʊv]　n. 丁香
cinnamon ['sɪnəmən]　n. 肉桂皮
dahurian angelica root　白芷
dried orange peel　陈皮
villous amomum fruit　砂仁
rinse [rɪns]　v.（用清水）冲洗,洗刷
pluck [plʌk]　v. 拔,摘,拔掉（死禽的毛）
blanch [blɑːntʃ]　v. 焯（把蔬菜等放在沸水中略微煮一会）
squeeze [skwiːz]　v. 挤,捏,压迫,压榨
pinch [pɪntʃ]　n. 捏,掐,拧

 Text A

　　From high-end restaurants to fast-food restaurants, poultry ingredients are popular and ideal for food-service operations because of versatility and low cost. At the same time, among diet-conscious people due to being lower in fat and cholesterol than other meat, the demand of chicken, ducks, geese and turkeys are increasing in popularity and availability.

　　The composition and structure of poultry flesh is almost the same as the meat of domestic animals, but poultry is not divided as many cuts as animal meat. Depending on the color of meat, free-range or farm-raised, chicken and turkey are composed of two parts: light meat and dark meat. So-called light meat, it usually refers to breast and wings which feature being less fat and connective tissue and being cooked faster and more easily. On the contrary, because of more fat and more connective tissue, the legs (drumsticks and thighs) which take longer to cook are called dark meat. Duck, goose and squab have all dark meat, so it is common to use different cooking methods to cook chicken and other poultry items.

In China, lots of chefs like to cook the whole birds by simmering, stewing and braising, while western chefs prefer cooking poultry parts due to the different cooking characteristics of each part.

Fresh poultry is extremely perishable and should be cooked within 24 hours. Frozen poultry should be stored at −18 ℃ or lower until ready to thaw.

The cooking methods of poultry are similar to the meat of domestic animals because their muscle tissue basically has the same structure. It is especially stressed that when western chefs cook the roasted or baked stuffed chicken or turkey, they often bake the stuffing separately, and then put the stuffing into the gutted chicken or turkey in order to get better results. The stuffing includes starch (rice or bread), aromatic vegetables (celery or onion), fat, stock, seasoning, eggs, sausages, nuts, fruits, and so on.

How to make broiled chicken

A: May I take you order, sir?

B: Yes, I'd like to have broiled chicken.

A: Sure, sir. But it takes quite a while to prepare this dish. Be patient, please.

B: How do you prepare it?

A: First of all, split the chicken in half and brush the chicken on both sides with melted butter. Season with salt and pepper.

B: What should you do then?

A: Place the chicken side down on the broiler rack. Broil at moderately low heat until the chicken is half cooked and well browned on one side.

B: And then?

A: Turn the chicken over. Continue to broil until the chicken is done and well browned on second side.

B: What is next?

A: Remove from the broiler. Place half a chicken on plate with skin side up and serve.

B: Sounds not complex. I will have a try this evening.

How to make "Dezhou-style Fried & Simmered Chicken"

Name of the dish: Dezhou-style Fried & Simmered Chicken(德州扒鸡)

Ingredients: dressed chicken(killed, gutted and stripped feather off)

Seasonings: 5 grams of black pepper   15 grams of ginger

15 grams of star anise   15 grams of cinnamon

5 grams of nutmeg

5 grams of fennel

15 grams of cloves

15 grams of dahurian angelica root

5 grams of dried orange peel

3 grams of villous amomum fruit

All these seasonings are packed into a linen-bag for cooking 10 dressed chickens.

Condiments: soy sauce, honey, salt

Cooking Procedure:

1. Rinse the 10 dressed chickens well and drain them.

2. Make sauce with honey and water and stir well. Color the dressed chickens evenly and cool them naturally.

3. Pour edible oil in a big wok and heat over high heat up to 210 ℃, and put the colored chickens in and fry for 2 minutes until they become gold-yellow and drain.

4. Put the fried chickens and the seasonings (the linen-bag) in a wok and pour fresh water to submerge the chickens, and add soy sauce and salt. Firstly cook by high heat until the water is boiling, and then simmer them over low heat for 3 to 4 hours.

5. Drain the cooked chickens well and dish up.

Exercises

Activity Ⅰ What kind of poultry can you see in the pictures? Please write them down, and then translate them into Chinese.

1. _____   2. _____   3. _____

4. _____   5. _____   6. _____

7. _____   8. _____   9. _____

Activity Ⅱ　Fill in the blanks with verbs.

1. _____ the cock　给公鸡拔毛
2. _____ the spicy duck neck into _____　把麻辣鸭脖切段
3. _____ the mallard　把鸭子洗净
4. _____ the turkey　取出火鸡的内脏
5. _____ the roast duck　把烤鸭切片
6. _____ the goose　炖鹅
7. _____ the chicken breast　把鸡胸肉焯水
8. _____ the stock into the wok　把高汤倒入锅内
9. _____ the soup over low heat　低温慢火煮汤
10. _____ the wings until they become brown　煎鸡翅直到颜色金黄
11. _____ a slice of lemon in the soup　在汤里挤一片柠檬汁
12. _____ a pinch of salt　加点盐

Activity Ⅲ　Translation.

奶油炖羊羔肉		
所需工具及设备	汤锅、平底锅、烤箱、擀面杖	
原料	数量/重量	步骤
羊骨	300克	
培根	100克	
胡萝卜	1根	
洋葱	1个	
大蒜	1瓣	香草夹层：
胡椒碎	1茶匙	将所有原料混合,用擀面杖擀平,包上保鲜膜冷藏备用
基础牛汤	500毫升	
基础鸡汤	750毫升	1.锅里加油,加入切好的培根炒出油,加入切碎的羊骨、胡萝卜、洋葱、大蒜炒熟。加入基础汤、香料束,加盐调味,小火煮90分钟,加入胡椒碎煮30分钟,沥干备用
香料束	1束	
柠檬蜜饯	100克	
羊里脊	800克	2.锅里加油,将羊肉煎上色,小火再煎4分钟,冷却后将香草夹层放到羊肉上于烤箱烤2～3分钟
雪莉酒醋	少许	
香草夹层：		3.同一个锅里加入少许雪莉酒醋、胡椒碎及少许骨头汤,煮至一半时加入柠檬蜜饯及黄油
黄油	100克	
面包糠	100克	4.成品羊肉切成3段,竖着排列
杏仁粉	50克	
面粉	20克	
大蒜	1瓣	
香料碎(法香,龙蒿草)	适量	
注意事项		

Extended Materials
猪肉各部位中英文名称

猪肉　pork	连骨头的猪排　pork chop
猪排　chop	卷好的腰部瘦肉　rolled pork loin
五花肉　streaky pork	卷好的腰部瘦肉连带皮　rolled pork belly
肥肉　fatty meat	做香肠的绞肉　pork sausage meat
瘦肉　lean meat	熏肉　smoked bacon
前腿　foreleg	小里脊肉　pork fillet
后腿　hind leg	带骨的瘦肉　spare rib pork chop
猪肝　pig liver	小排骨肉　spare rib of pork
猪脚　pig feet	肋骨　pork rib
猪腰　pig kidney	猪油　lard
猪心　pig heart	蹄髈　hock
猪肚　pig bag	中间带骨的腿肉　casserole pork
没骨头的猪排　pork steak	有骨的大块肉　joint

羊肉各部位中英文名称

羊外脊（沿颈后到腹腔脊骨背侧肉）boneless loin	羊霖（大腿前与腰窝处肉）　thick flank
羊里脊（沿脊背到腹腔脊骨内侧条肉）tenderloin	针扒（后臀肉）　topside
后腿部（肉质略老,瘦）　rump	颈肉　neck
尾龙扒（腰臀肉,去盖）　cap off	去骨羊肩肉（肩背肉）　shoulder fillet
烩扒（磨裆肉）　silverside	去骨羊腩（腹肌,五花肉）　flap-boneless
	腱子（前后小腿肉,瘦）　shank

鸡肉各部位中英文名称

大鸡腿　fresh chicken leg	小鸡腿　chicken drumstick
鸡胸肉　fresh chicken breast	鸡翅膀　chicken wing

牛肉各部位中英文名称

小块的瘦肉　stewing beef	可煎食的大片牛排　frying steak
牛肉块加牛腰　steak & kidney	牛绞肉　minced beef

大块牛排　rump steak	蜂巢状牛肚　beef honeycomb tripe
牛尾　ox-tail	牛肚块　beef tripe piece
牛心　ox-heart	白牛肚　beef best thick seam
牛舌　ox-tongue	牛展（小腿肉）　beef shin
带骨的腿肉　Barnsley chop	牛前（颈背部肉）　beef crop
肩肉　shoulder chop	牛胸（胸部肉）　beef brisket
腰上的牛排肉　porter house steak	西冷（腰部肉）　beef sirloin
头肩肉筋　chuck steak	牛柳（里脊肉）　beef tenderloin
拍打过的牛排　tenderized steak	牛腩（腹部肉）　beef thin flank
牛肠　beef roll	针扒（股内肉）　beef top-side
牛筋　beef cow-hell	尾龙扒（臀肉）　beef rump

Unit 6

Fish

Part A Understanding Fish and Shellfish

 Learning Goals

You will be able to:
1. know the classification of fish;
2. tell the name of the common fish of freshwater and saltwater;
3. know how to store fish.

 Vocabulary Assistance

edible ['edəbl] adj. 可食用的
nutritious [nju'trɪʃəs] adj. 有营养的,营养丰富的
refrigeration [rɪˌfrɪdʒə'reɪʃn] n. 冷冻,制冷
technology [tek'nɒlədʒi] n. 科技,工艺
fin fish 鱼类
shellfish ['ʃelfɪʃ] n. 贝类
skeleton ['skelɪtn] n. 骨骼,骨架
mackerel ['mækrəl] n. 鲭(鱼)
Spanish mackerel 鲅鱼
sardine [ˌsɑː'diːn] n. 沙丁鱼
grouper ['gruːpə(r)] n. 石斑鱼
pomfret ['pɒmfrɪt] n. 鲳鱼
yellow croaker 黄花鱼
flatfish ['flætfɪʃ] n. 比目鱼
salmon ['sæmən] n. 鲑,三文鱼
ribbon fish 带鱼

ocean perch 鲈鱼
tuna ['tjuːnə] n. 金枪鱼
cod [kɒd] n. 鳕鱼
carp [kɑːp] n. 鲤鱼
black carp 青鱼
grass carp 草鱼
silver carp 鲢鱼
variegated carp 鳙鱼
catfish ['kætfɪʃ] n. 鲇鱼
black fish 黑鱼
crucian 鲫鱼
trout [traʊt] n. 鳟鱼
tissue ['tɪʃuː] n. (人、动植物的)组织,(尤指用作手帕的)纸巾
excessive [ɪk'sesɪv] adj. 过分的,过度的
thaw [θɔː] v. (结冰后)解冻,融化

 Text

The edible flesh of fish is nutritious which consists of water, proteins, fat and small amounts of minerals, vitamins and other substances. Nowadays, thanks to modern refrigeration and freezing technology, fish products are enjoyed not in limited areas (along the seacoast or around lakes and rivers) but much more widely. As far as cooking is concerned, fish products fall into two categories: fin fish (fish for short) that refers to fish with fins and internal skeletons; shellfish which means fish have external shells but no internal bone structure.

As for fin fish, the significant difference in flavor between freshwater fish and saltwater fish is that the latter have more salt in their flesh. The common saltwater fish are (Spanish) mackerel, sardine, grouper, pomfret, yellow croaker, sole, salmon, ribbon fish, ocean perch (sea bass), tuna, cod, and so on. In contrast, carp, black carp, grass carp, silver carp, variegated carp, catfish, black fish, crucian and trout are all freshwater fish.

Compared with meat, fin fish has very little connective tissue, is very delicate and is easily overcooked. Lean fish, such as flounder, cod bass and perch, is often served with sauces to enhance moistness and give richness. But flat fish, such as salmon, tuna, trout and mackerel, is usually cooked with dry-heat methods to help to get rid of excessive oiliness.

Waiter: Good evening, sir. Are you ready to order?

Guest: No, I have no idea now. Would you like to recommend some specialty to me?

Waiter: My pleasure, sir. As you know, seafood is popular in Qingdao city. Saltwater fish are delivered to us each day. We also offer freshwater fish.

Guest: Sounds great. I have a poor appetite today, so I'd like to take some light food.

Waiter: OK, sir. Would you care for the steamed perch, the chef's recommendation today?

Guest: Good, I'll take it.

Activity I What seafood can you see in the pictures? Please write them down, and then translate them into English.

1._____ 2._____ 3._____

4._____ 5._____ 6._____

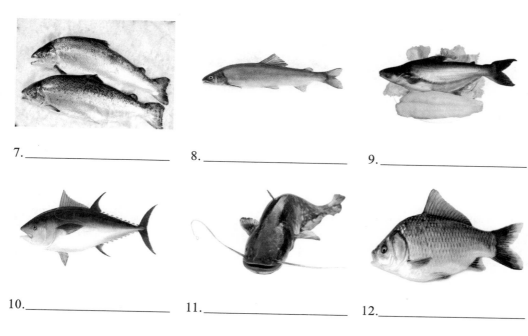

7. _____ 8. _____ 9. _____

10. _____ 11. _____ 12. _____

Activity Ⅱ　Classify these fish in the above pictures.

Freshwater fish：
_____

Saltwater fish：
_____

Focus on Language

 Listen to the following passage and fill in the blanks.

Fish may be stored 1 to 2 days in the 1._____ after your purchase, but it is not the best way to store it. The most 2._____ method to store fish is to put it on the crushed ice to maintain the temperature of －2 ℃ to 0 ℃. If you want to keep it 3._____, you must 4._____ it in moisture-proof 5._____ or plastic bag and freeze it or cook and refrigerate it for 6._____ use. So it is important to store it carefully and use it quickly because fish can 7._____ easily. Frozen raw fish had better 8._____ in refrigerator and not at room temperature for 18 to 36 hours depending on 9._____. Remember, never 10._____.

Translation

1. 这道菜真好吃，你能告诉我是什么鱼做的吗？
2. 服务员：女士，请问您想吃点什么？
   顾客：我今天胃口不好，想吃点清淡的。
3. 服务员：您不想尝尝我们的特色菜吗？我推荐粉蒸鸡(steamed chicken with rice flour)。
   顾客：我想尝一下本地的海鲜菜，软炸(soft-dried)虾仁怎么样？
4. 我想要一份清蒸鲈鱼，另外配一份时令蔬菜(vegetables in season)。

外教有声

Part B Shellfish

Learning Goals

You will be able to:
1. know the classification of shellfish;
2. tell the name of the common shellfish;
3. know how to store fish.

Vocabulary Assistance

mollusk   n. 软体动物
bivalve ['baɪvælv]   n. 双壳软体动物
univalve   n. 单壳软体动物
cephalopod ['sefəlɒpɒd]   n. 头足动物（如章鱼和乌贼）
crustacean [krʌ'steɪʃn]   n. 甲壳纲动物（如螃蟹、龙虾和虾）
oyster ['ɔɪstə(r)]   n. 牡蛎,蚝
clam [klæm]   n. 蛤,蛤蜊
mussel ['mʌsl]   n. 贻贝
scallop ['skɒləp]   n. 扇贝

squid [skwɪd]   n. 乌贼,鱿鱼
octopus ['ɒktəpəs]   n. 章鱼
lobster ['lɒbstə(r)]   n. 龙虾,(供食用的)龙虾肉
shrimp [ʃrɪmp]   n. 虾,小虾
prawn [prɔːn]   n. 对虾,大虾
crab [kræb]   n. 蟹,螃蟹,蟹肉
simmer ['sɪmə(r)]   v. 用文火炖,煨
broil [brɔɪl]   v. 烤,焙（肉或鱼）
leftover ['leftəʊvə(r)]   n. 吃剩的食物,残羹剩饭

Text

Compared with fin fish, shellfish is featured by their harder outer shells and their lack of the backbone or internal skeleton.

Generally speaking, there are two kinds of shellfish. The first one refers to mollusks which are soft sea animals that are divided into three main categories: bivalve, univalve and cephalopod. The second one is named crustacean.

In the market, oysters, clams, mussels and scallops are common mollusks which have a distinguished feature by a pair of hinged shells. Abalone and conch are called univalves because they have a single shell. So-called cephalopod means "head-foot", referring to the fact that these sea animals have tentacles or "leg", attached to the head and surrounding the mouth. Squid and octopus are often seen in the kitchen.

The most popular crustaceans in kitchens all over the world are lobsters, shrimps and crabs. The term "prawn" is sometimes used for large shrimps. As for crabs, there are many kinds of them in the commercial kitchens, and among them, the expensive kinds are Alaskan king crab and Alaskan snow crab.

The most common cooking methods for shellfish are steaming, simmering and broiling. When cooking them, please pay attention to duration and degree of heating.

Dialogue

Tom: Hi, Mom, what is for dinner?

Mom: Would you like some rice? We have some leftover from last night. Is it OK if I make fried rice with eggs?

Tom: Fried rice is OK, but can we have something freshly-made tonight?

Mom: All right. How about steamed crabs and boiled shrimps?

Tom: Terrific. I love seafood. Can I help you?

Mom: I don't think so.

Tom: It's fun to cook.

Mom: Oh, really? Then help me to take the crabs and shrimps out of the refrigerator and wash them. I just bought them today. I will make the fried rice.

Tom: No problem, Mom. Today I learn to cook.

(After a while)

Tom: It's done, Mom. What's next?

Mom: Hmm... It only takes 10 minutes to get the crabs done in the rice cooker. Then you can boil the shrimps. Put a little cooking wine, some sliced ginger and shrimps into the boiling water, 5 minutes later, you can serve them.

Tom: Thanks, mom. It is so easy.

Exercises

Activity I What seafood can you see in the pictures? Please write them down, and then translate them into English.

1. _____ 2. _____ 3. _____

4. _____ 5. _____ 6. _____

7. _____ 8. _____ 9. _____

Activity Ⅱ　Fill in the blanks with verbs.

1. _____ the black fish　把黑鱼的鳞刮掉
2. _____ the fins of the perch　剪掉鲈鱼的鳍
3. _____ the crucian　取出鲫鱼的内脏
4. _____ the crab　蒸螃蟹
5. _____ the clams　翻炒蛤蜊
6. _____ the lobster　烤龙虾
7. _____ the shrimps　煮虾
8. _____ the yellow croaker　炸黄花鱼
9. _____ the scallop salad　调拌扇贝沙拉
10. _____ sea-cucumber in soy sauce　红烧海参
11. _____ the ribbon fish　去掉带鱼头部
12. _____ the shrimp　去虾线
13. _____ the fish　剔除鱼刺
14. _____ the sole　煎比目鱼
15. _____ the salmon roll　熏三文鱼卷

Activity Ⅲ　Translation.

香煎鳕鱼配薏米

主料:鳕鱼 300 克、薏米 200 克。

辅料:海虹水、罗勒 5 克、意大利香菜 5 克等。

调料:盐 5 克、色拉油、黑胡椒 5 克等。

制作过程:

1. 用海虹水煮熟薏米,鳕鱼切成小段,罗勒切丝。
2. 锅烧热到 300 ℃撒上少许色拉油,鳕鱼皮朝下煎 3～4 分钟。
3. 用铁铲将鳕鱼翻面继续煎 3～4 分钟。
4. 将鳕鱼撒上盐、胡椒,刷上橄榄油后放入 190 ℃的烤箱烤 7 分钟。
5. 盘子用海虹水煮过的薏米铺底,撒上意大利香菜、罗勒丝。
6. 鳕鱼铺在上面,平面向上,撒上盐,用食用花装饰。

Part C How to Cook a Seafood Dish

  Learning Goals

You will be able to:

1. ask and answer for advice in making a fish;
2. tell the preparation of materials;
3. tell the process of making seafood dishes.

  Vocabulary Assistance

bake [beɪk]   v. 烘烤，焙
sesame ['sesəmi]   n. 芝麻
breadcrumbs ['bredkrʌmz]   n. 面包屑，面包碎
gut [gʌt]   n. 消化道，肠道，（尤指动物的）内脏
   v. 取出……的内脏（以便烹饪）

sprinkle ['sprɪŋkl]   v. 撒，洒，把……撒（或洒）在……上
garnish ['gɑːnɪʃ]   v. （用菜）为（食物）加装饰，加饰菜于
streaky pork ['striːki pɔːk]   五花肉
scale [skeɪl]   v. 去鳞

  Dialogue

Baked fish with sesame seeds

Ingredients:

onions, vegetable oil, sesame seeds, water, chopped garlic, salt, lemon juice, red pepper, breadcrumbs, fish

Dialogue:

Commis: What shall I do with the fish?

Chef: Scale it, cut off the fins, and then gut it.

Commis: Gut it?

Chef: Yeah, use the fish scissors to cut open the stomach of the fish. Then take out the guts.

Commis: OK. What is next?

Chef: Cook the onions.

Commis: I already did.

Chef: Good. Mix the sesame seeds, water, garlic, salt, lemon juice and red pepper with the spoon.

Commis: All right. What is next?

Chef: Sprinkle the baking dish with breadcrumbs and parsley.

Commis: And then put the fish in the baking dish?

Chef: Yes, pour the sesame seeds and onions over the fish. Light the oven and cook the fish at

外教有声

400 °F.

Commis: For how long?

Chef: For twenty or twenty-five minutes. When it is cooked, garnish the fish with parsley and olives.

Steamed Perch

Ingredients: perch (500 g), ginger, scallion, streaky pork, Chinese black mushroom, black soy sauce, salt, cooking wine, gourmet powder, black pepper powder.

Mince-plus period:

Firstly, scale the perch, remove gills and gut it, cut off the tails.

Secondly, carve slightly along the perch's back.

Thirdly, slice the streaky pork.

Fourthly, remove the seeds of Chinese black mushroom.

Fifthly, slice the ginger and cut the scallion into small chunks.

Cooking period:

1. Put some cooking wine, black pepper powder and salt on fish, then rub it.

2. Put sliced streaky pork, sliced ginger, chunked scallion, Chinese black mushroom on fish and stuff some in fish.

3. Put fish into the steamer, and then steam it for 7-8 minutes.

4. Take off pork, ginger, scallion and Chinese black mushroom.

5. Pour the original sauce into the pot, and add some soy sauce with a little cooking starch and boiled oil, then boil it for 1-2 minutes.

6. Pour the sauce onto the fish.

Sum-up:

Bill of material: perch, streak pork, ginger, scallion, Chinese black mushroom, and so on.

Cooking keys:

1. Carve slightly along the perch's back.

2. Steaming time should be 7 to 8 minutes.

Feature:

Being deliberate, tender, smooth, fresh and salty.

Activity I

Role Playing A

Commis: What shall I do with _____? (the fish/the shrimp)

Chef: _____? (scale it/cut off the fins/gut it/dehead it/devein it)

Commis: I already _____. (did/have)

Role Playing B

Chef: Sprinkle the baking dish with _____? (breadcrumbs/parsley/cauliflower/onion)

Commis: And then put _____ in the baking dish? (the fish/salmon/tuna)

Activity II Translate the following sentences into English.

洛克菲勒风格焗生蚝

主料:本地生蚝、混合面包糠。

调料:盐、胡椒、黄油、帕玛森芝士、意大利芹碎、柠檬皮碎、培根碎。

装饰:帕尔玛火腿片、迷迭香、柠檬角。

制作过程:

1. 将生蚝洗净。
2. 煮一锅开水。
3. 将生蚝倒入开水中烫至开口。
4. 将生蚝过凉取出洗净。
5. 将凹进去的半壳洗干净。
6. 将洗好的生蚝肉放回干净的壳内备用。
7. 出菜前将生蚝肉烫热。
8. 撒上混合面包糠烤熟、上色即可。

混合面包糠:将所有的调料及面包糠拌匀即可。

备注:视季节、生蚝大小而定数量。

Extended materials

常见鱼类的中英文名称

albacore　花鱼
anchovy　凤尾鱼,鳀,银鱼柳
bass　鲈鱼,河鲈
white bass　白鲈
black fish　黑鱼
bluefish/scad　竹荚鱼类,美洲大西洋沿岸的青鱼
bream　鳊鱼
striped bream　斑纹鳊鱼
buffalo fish　牛鱼(一种分布于北美地区的黑色大淡水鱼)

butterfish　鲳鱼
carp　鲤鱼
grass carp　草鱼,鲩鱼
silver carp　鲢鱼
catfish　鲇鱼
codfish　鳕鱼类的统称
black cod/sablefish　黑鳕鱼,裸盖鱼
coalfish　黑鳕鱼
cod/lingcod/ling　鳕鱼
cod cheek　一种上等昂贵的鳕鱼

croaker/drum 鼓鱼
corvine/yellow croaker 石首鱼,黄花鱼
spotfin croaker 斑鳍鼓鱼
redfish/red drum 雄鲑,红鼓鱼
white sea bass 太平洋犬牙石首鱼
kingfish 无鳔石首鱼
spot croaker 斑点鼓鱼
crucian 鲫鱼
dace 鲮鱼,鲦鱼,雅罗鱼
dogfish/Cape shark 角鲨,狗鱼
dory 海鲂
eel 鳗鱼,鳗鲡
American eel 美洲鳗鱼
conger eel/marine eel 海鳗
eel ladder 梯状鳗鱼
elver/baby eel 鳗苗
European eel 欧洲鳗鱼
garden eel 花园鳗鱼
lamprey 七鳃鳗
moray eel 欧洲海鳗
short-finned eel 短鳍鳗鱼
snowflake moray 欧洲雪花海鳗
spiny eel 多刺的鳗鱼
unagi 河鳗
freshwater eel 白鳝
yellow eel 鳝鱼,黄鳝
flatfish/flounder 比目鱼,偏口鱼,龙利鱼
turbot 大菱鲆,大比目鱼(比目鱼中的精品鱼种)
gar/garfish/needle fish 雀鳝,长嘴硬鳞鱼,针鱼
globefish 河豚
grouper 石斑鱼
jewfish 海鲈鱼,暖海鱼(石斑鱼的一种)
striped bass 条纹鲈鱼
gurnard/sea-robin 海鲂
herring 鲱鱼
whitebait/young herring 银鱼,属于鲱的幼体
jerk filefish 马面鱼

mackerel 鲭(鱼)
mackerel pike 秋刀鱼
Spanish mackerel 马鲛鱼,鲅鱼
loach 泥鳅
mandarin fish 鳜鱼
marlin 枪鱼(包括太平洋青枪鱼 Pacific blue marlin,适合做寿司)
kajiki (日语)枪鱼(适合做寿司)
milkfish 遮目鱼,虱目鱼
monkfish 安康鱼,扁鲨,琵琶鱼,华脐鱼
anglerfish/bellyfish 琵琶鱼,华脐鱼,安康鱼(扁鲨的一种)
frogfish 蟾鱼科(安康鱼的一种)
sea-devil 扁鲨(安康鱼的别称)
mullet 胭脂鱼,鲻鱼
black mullet/striped mullet 黑鲻鱼,斑点鲻鱼
golden grey mullet/grey mullet 灰鲻鱼
parrot fish 鲻鱼的一种
red mullet 红鲻鱼
thin lipped mullet 细唇鲻鱼
white mullet 白鲻鱼
perch 河鲈,鲈鱼
Lake Victoria perch 维多利亚湖鲈鱼
Nile perch 尼罗河鲈鱼
ocean perch/sea perch 海鲈鱼
walleye/wall-eyed pike 白斑鱼
pike 梭子鱼
pomfret 鲳鱼,银鲳
pompano/black pomfret 黑鲳
porgy 鲷鱼,棘鬣鱼,大眼鱼,大西洋鲷,钉头鱼
scup/fair maid 变色窄牙鲷,美女鲷
sea bream 海鲷,加吉鱼
red porgy 红海鲷
sheepshead porgy 羊头鲷(鲷鱼的一种)
shad porgy 类似于西鲱的鲷鱼
whitebone porgy 白骨鲷鱼
jolthead porgy 笨头鲷鱼
ribbon fish/hairtail 带鱼

rockfish　岩鱼
black sea bass　巨大硬鳞鱼
salmon　三文鱼,鲑鱼,大马哈鱼
chinook salmon/king salmon/spring salmon
　　大鳞三文鱼,大马哈鱼
sardine/pilchard　沙丁鱼
sculpin　松江鱼
sea bass　海鲈鱼
California sea bass　加利福尼亚黑鲈
sheepshead　羊头鱼,红鲈
cabrilla　热带海水鲈鱼
shad　西鲱,美洲河鲱
shark　鲨鱼
swellfish　河豚
swordfish/sailfish　旗鱼,剑鱼
terch　鳙鱼
tilapia　罗非鱼,吴郭鱼,非洲鲫鱼
tilefish/ocean whitefish　方头鱼
trout　鲑鱼,鳟鱼
rainbow trout/steelhead　虹鳟鱼

sea trout　海鳟鱼
tuna　金枪鱼,吞拿鱼(适合做日餐的寿司和生鱼片)
albacore　长鳍金枪鱼
bluefin　蓝鳍金枪鱼(做生鱼片的精品鱼肉)
blackfin　黑鳍白鲑,黑鳍金枪鱼
skate wing　老板鱼
smelt　胡瓜鱼
snakehead　黑鱼
sole　鲽鱼,板鱼(比目鱼的一种,包括gray sole、petrale sole、English sole 和 Rex sole)
yellow fin sole　黄鳍鲽鱼
sturgeon　鲟鱼
mandarin sturgeon　中华鲟
sunfish　翻车鱼(一种淡水小鱼)
yellowfin　黄鳍金枪鱼
wrasse　濑鱼
wreckfish　一种常聚集于沉船周围的鱼

Unit 7

Fruits and Vegetables

Part A　Fruits

Learning Goals

You will be able to:
1. know the classification of fruits;
2. tell the name of the common fruits;
3. know how to cook fruits.

Vocabulary Assistance

fruit [fruːt]　n. 水果
apple [ˈæpl]　n. 苹果
banana [bəˈnɑːnə]　n. 香蕉
pear [peə(r)]　n. 梨子
peach [piːtʃ]　n. 桃子
orange [ˈɒrɪndʒ]　n. 橘子, 橙子
watermelon [ˈwɔːtəmelən]　n. 西瓜
melon [ˈmelən]　n. 甜瓜
grape [greɪp]　n. 葡萄
grapefruit [ˈgreɪpfruːt]　n. 西柚, 葡萄柚
lychee [ˈlaɪtʃiː]　n. 荔枝
pineapple [ˈpaɪnæpl]　n. 菠萝
mango [ˈmæŋgəʊ]　n. 芒果
cherry [ˈtʃeri]　n. 樱桃

plum [plʌm]　n. 李子, 梅子
apricot [ˈeɪprɪkɒt]　n. 杏
lemon [ˈlemən]　n. 柠檬
lime [laɪm]　n. 青柠
Chinese date [deɪt]　枣子
coconut [ˈkəʊkənʌt]　n. 椰子
dragon fruit [ˈdrægən fruːt]　火龙果
kiwi fruit　奇异果
avocado [ˌævəˈkɑːdəʊ]　n. 牛油果, 鳄梨
blueberry [ˈbluːbəri]　n. 蓝莓
strawberry [ˈstrɔːbəri]　n. 草莓
raspberry [ˈrɑːzbəri]　n. 树莓, 覆盆子
blackberry [ˈblækbəri]　n. 黑莓
cranberry [ˈkrænbəri]　n. 蔓越莓

Text

　　Fruit is a kind of edible plant fruit with juicy, sweet and sour taste. It is not only rich in vitamin nutrition, but also promote digestion. It contains more sugar and vitamins, and also contains many kinds of special substances with biological activity, so it has high nutritional value and health function.

　　Fruit raw materials have a variety of colors, which are widely used in Western food. They can be used to make salad and dessert, we can also use fruits in cocktails and decoration of the dishes, especially in American cuisine. We can eat fruit in different ways, eat directly or after cooking.

 Dialogue

Peter: What is your favorite fruit?

Jane: My favorite fruit is grape.

Peter: Oh, really?

Jane: Yes.

Peter: What kind of grape do you like? Green grapes or purple grapes?

Jane: I like purple grapes. How about you, what is your favorite fruit?

Peter: I like the avocado.

Jane: Oh, I have not eaten the avocado before. How's the taste?

Peter: The avocado is delicious, especially with honey. We can mix diced avocado with milk and honey.

Jane: OK, I will try next time.

 Exercises

Activity What fruits can you see in the pictures? Please write them down, then translate them into English.

1. _____   2. _____   3. _____

4. _____   5. _____   6. _____

7. _____   8. _____   9. _____

Focus on Language

🎧 **Listen to the following recipe and fill in the blanks.**

Baked Apple
Ingredients:
4 large good baking apples
1/4 cup of 1._____
1 teaspoon of cinnamon
1/4 cup of chopped pecans
1/4 cup of currants or 2._____
1 tbsp of 3._____
3/4 cup of 4._____

Cooking steps:
1. 5._____ oven to 375 ℉. Wash apples. 6._____ cores to 1/2 inch of the bottom of the apples.
2. In a small bowl, 7._____ the sugar, cinnamon, currants/raisins, and pecans. Place apples in a 8-inch-by-8-inch square baking pan. 8._____ each apple with this mixture, 9._____ a dot of butter.
3. Add boiling water to the 10._____. Bake 30-40 minutes, until 11._____, but not mushy. 12._____ the oven and baste the apples several times with the pan juices.
4. Serve warm with vanilla ice-cream on the side.

Part B Vegetables

Learning Goals

You will be able to:
1. know the classification of vegetables;
2. tell the name of the common vegetables;
3. know how to cook vegetables.

Vocabulary Assistance

vegetable ['vedʒtəbl]   n. 蔬菜
potato [pə'teɪtəʊ]   n. 土豆,马铃薯
tomato [tə'mɑːtəʊ]   n. 西红柿,番茄
onion ['ʌnjən]   n. 洋葱
red onion [red 'ʌnjən]   红皮洋葱
carrot ['kærət]   n. 胡萝卜
cucumber ['kjuːkʌmbə(r)]   n. 黄瓜

corn [kɔːn]   n. 玉米
corncob ['kɔːnkɒb]   n. 玉米芯
mushroom ['mʌʃrʊm]   n. 蘑菇
pepper ['pepə(r)]   n. 辣椒
bell pepper [bel 'pepə(r)]   甜椒,灯笼椒
celery ['seləri]   n. 芹菜
lettuce ['letɪs]   n. 生菜

cabbage ['kæbɪdʒ]　n. 卷心菜
Chinese cabbage　大白菜
purple cabbage　紫甘蓝
purple ['pɜːpl]　n. 紫色 adj. 紫色的
watercress ['wɔːtəkres]　n. 西洋菜
endive ['endaɪv]　n. 苦苣，菊苣
broccoli ['brɒkəli]　n. 西蓝花
cauliflower ['kɒlɪflaʊə(r)]　n. 花椰菜
asparagus [ə'spærəgəs]　n. 芦笋
beetroot ['biːtruːt]　n. 红菜根，甜菜
turnip ['tɜːnɪp]　n. 芜菁
pumpkin ['pʌmpkɪn]　n. 南瓜
winter squash [skwɒʃ]　节瓜
zucchini [zuˈkiːni]　n. 西葫芦
bean [biːn]　n. 豆，菜豆，豆荚
broad bean　蚕豆
pea [piː]　n. 豌豆

string bean [ˌstrɪŋ 'biːn]　菜豆
okra ['əʊkrə]　n. 秋葵
leaf [liːf]　n. 叶子
eggplant ['egplɑːnt]　n. 茄子
head [hed]　n. 头，（植物的茎梗顶端的）头状叶丛
parsley ['pɑːsli]　n. 欧芹
garlic ['gɑːlɪk]　n. 大蒜
ginger ['dʒɪndʒə(r)]　n. 生姜
sprinkle ['sprɪŋkl]　v. 撒
stir [stɜː(r)]　v. 搅拌
occasionally [ə'keɪʒnəli]　adv. 偶尔，偶然
slice up　将……切成薄片
broth [brɒθ]　n. （加入蔬菜的）肉汤，鱼汤
thyme [taɪm]　n. 百里香
simmer ['sɪmə(r)]　v. 炖

Text

There are many kinds of vegetables, which can be divided into leaf vegetables, root vegetables, stem vegetables, melon vegetables, bean vegetables, cauliflower, and so on.

Leaf vegetables also called vegetable greens, leafy greens, or salad greens, sometimes accompanied by tender petioles and shoots. Although they come from a very wide variety of plants, most share a great deal with other leaf vegetables in nutrition and cooking methods.

The edible part of root vegetables is the root of plants. The common raw materials in Western food are carrot, turnip, red cabbage head, etc, which are usually grown in the soil. They are washed and peeled before cooking, which is suitable for a variety of cooking methods.

The edible part of the stem vegetables is the stem of plants. The common raw materials in Western food are onion, garlic, etc.

Melon vegetables are vegetables that take the fruits of plants as cooking materials, for example, pumpkin, cucumber, and so on.

For the bean vegetables, the edible part is the seed part of the plant, such as the okra.

Cauliflower is a kind of vegetable which takes the flower as the edible part, such as broccoli, cauliflower, and so on.

Dialogue

Commis: What shall I do?
Chef: Slice up forty onions.
Commis: I did it.
Chef: Cook the onions in butter over low heat.

外教有声

Commis: Yes, over low heat and stir occasionally. For how long?

Chef: Twenty or thirty minutes.

(After twenty-five minutes)

Commis: What should I do next?

Chef: Add the beef broth, water, bay leaves, pepper and thyme.

Commis: I will heat it to boiling.

Chef: Fine, then reduce the heat. Cover and simmer.

Commis: And in the meantime?

Chef: Toast the French bread.

(After fifteen minutes)

Commis: I put the toasted bread into bowls and poured onion soup on top.

Chef: OK. Now sprinkle Parmesan cheese on top at last.

Exercises

Activity What vegetables can you see in the pictures? Please write them down, and then translate them into English.

1._____ 2._____ 3._____

4._____ 5._____ 6._____

7._____ 8._____ 9._____

Focus on Language

Listen to the following sentences and fill in the blanks.

1. _____ is the heart of most salads.
2. Use the outer _____ for decoration.
3. Cut up the inner leaves into the _____ pieces.
4. Refrigerate lettuce to keep it _____.
5. Use _____ color to make salads look nice.
6. Add _____ to the salad seconds before serving.
7. Serve salad on a cold _____.

Part C Processing Methods of Fruits and Vegetables

Learning Goals

You will be able to:
1. know the washing method of fruits and vegetables;
2. know the peeling and cutting method of fruits and vegetables;
3. know the basic principles of vegetable cooking.

Vocabulary Assistance

peel [piːl] v. 剥（水果、蔬菜等的）皮，去皮，剥掉
scrub [skrʌb] v. 擦洗
drain [dreɪn] v. 沥干
soak [səʊk] v. 浸泡
salted [ˈsɔːltɪd] v. 在食物中放盐，用盐腌制
brown [braʊn] v. 使变成棕色
overcook [ˌəʊvəˈkʊk] v. 煮得过熟
reheat [ˌriːˈhiːt] v. 重新加热
baking soda   碳酸氢钠，小苏打
even [ˈiːvn] adj. 均匀的，平稳的，相等的，均等的
strong-flavored   味道浓烈的
batch [bætʃ] n. 一批
boil [bɔɪl] v. 煮

crush [krʌʃ] v. 压碎
mix [mɪks] v. 混合
poach [pəʊtʃ] v. 水煮（如水煮荷包蛋）
plate [pleɪt] n. 盘子
dressing [ˈdresɪŋ] n. （拌制色拉用的）调料
bite [baɪt] v. 咬，叮
              n. 一口食物，少量食物
thoroughly [ˈθʌrəli] adv. 非常，极其，彻底，完全
leach out 溶出
eliminate [ɪˈlɪmɪneɪt] v. 消除
limp [lɪmp] adj. 柔软的，不直挺的
uniformly [ˈjuːnɪfɔːmli] adv. 均匀地，一致地

The washing method of fruits and vegetables

Washing:

1. Wash all vegetables thoroughly.

2. Root vegetables that are not peeled, such as potatoes for baking, should be scrubbed very well with a stiff vegetable brush.

3. Wash green leaf vegetables in several changes of cold water. Lift the greens from the water so the sand can sink to the bottom.

4. After washing, drain well and refrigerate lightly covered. The purpose of covering is to prevent drying, but covering too tightly cuts off air circulation. This can be a problem if the product is stored more than a day because mold is more likely to grow in a damp, closed space.

Soaking:

1. With a few exceptions, do not soak vegetables for long periods, flavor and nutrients will leach out.

2. Cabbage, broccoli and cauliflower may be soaked for 30 minutes in cold salted water to eliminate insects, if necessary.

3. Limp vegetables can be soaked briefly in cold water to restore crispness.

Peeling and cutting method of fruits and vegetables:

1. Peel most vegetables as thinly as possible. Many nutrients lie just under the skin.

2. Cut vegetables into uniform pieces for even cooking.

3. Peel fruits and vegetables as close to cooking time as possible to prevent drying and loss of vitamins through oxidation.

4. For machine paring, sort vegetables for evenness of size to minimize waste.

5. Treat vegetables that brown easily (potatoes, eggplant, sweet potatoes) with an acid, such as lemon juice until ready to use.

Basic principles of vegetable cooking

1. Don't overcook.

2. Cook as close to service time as possible, and in small quantities. Avoid holding for long periods on a steam table.

3. If the vegetable must be cooked ahead, undercook slightly and chill rapidly. Reheat at service time.

4. Never use baking soda with green vegetables.

5. Cut vegetables uniformly for even cooking.

6. Cook green vegetables and strong-flavored vegetables uncovered.

7. To preserve color, cook red and white vegetables in a slightly acid (not strongly acid) liquid. Cook green vegetables in a neutral liquid.

8. Do not mix a batch of freshly cooked vegetables with a batch of the same vegetables that was cooked earlier and kept hot in a steam table.

Dialogue A

Commis: What makes vegetables lose their vitamins?
Chef: Cutting them into small pieces.
Commis: What makes vegetables lose their color?
Chef: Overcooking them.
Commis: What is the best way to cook vegetables?
Chef: In small quantities in a pressure cooker.

Dialogue B

Commis: Shall I wash the cauliflower?
Chef: Yes, wash it well.
Commis: OK. Cauliflower is always dirty.
Chef: Soak the cauliflower in salt water at least twenty minutes.
Commis: How shall I cook the cauliflower?
Chef: Boil it in water with lemon juice.
Commis: The lemon juice will keep it white.
Chef: Yes.

Exercises

Activity   Fill in the blanks with verbs.

1. _____ potatoes   擦洗土豆
2. _____ the seeds of the pepper   去掉青椒的籽
3. _____ salad leaves   清洗沙拉叶子
4. _____ cauliflower   浸泡花椰菜
5. _____ pumpkin   去掉南瓜皮
6. _____ carrot   煮胡萝卜
7. _____ eggs   煮荷包蛋
8. _____ garlic   压碎大蒜
9. _____ cabbage   切开卷心菜
10. _____ onions   切碎洋葱

Focus on Language

Listen to the following dialogue and fill in the blanks.

Chef: 1._____ the tomatoes.
Cook: Shall I cut the 2._____?
Chef: Of course.
Cook: What should I do next?
Chef: Put the tomatoes in 3._____.
Cook: For how long?
Chef: For fifteen seconds.

Cook: And then?
Chef: Put them in 4._____ water.
Cook: Why?
Chef: Because we can 5._____ them easily and quickly.
Cook: OK. What will we use the 6._____ for?
Chef: Tomato 7._____.

Unit 8

Soup

Part A Classification of Soup

 Learning Goals

You will be able to:
1. know the definition of soup;
2. know the classification of soup.

 Vocabulary Assistance

ingredient [ɪnˈɡriːdiənt]　n. 成分,（尤指烹饪）原料
consommé [kənˈsɒmeɪ]　n. 清炖肉汤
stock [stɒk]　n. 现货,存货,库存;（尤指烹饪）高汤
extract [ˈekstrækt]　v. 提取,提炼
broth [brɒθ]　n. 肉汤
savory [ˈseɪvəri]　adj. 好吃的

puree [ˈpjʊəreɪ]　n. 酱,糊,泥
garnish [ˈɡɑːnɪʃ]　v. 为（食物）加装饰
　　　　　　　　　n.（食物上的）装饰菜
gravy [ˈɡreɪvi]　n. 肉汁
condiment [ˈkɒndɪmənt]　n. 调味料,作料
vegetable [ˈvedʒtəbl]　n. 蔬菜
seafood [ˈsiːfuːd]　n. 海鲜
chicken [ˈtʃɪkɪn]　n. 鸡,鸡肉

 Text

Soup is a food that is made by combining ingredients such as meat or vegetables in stock or hot/boiling water, until the flavor is extracted, forming a broth.

Classification　Generally speaking, soup is divided into 7 kinds in Western cuisine.

❶ Clear soup

Consommé, also called the consommé broth, in which beef, chicken and other spices, through boiling concentrated, seasoning and filtering steps, are made well, so that the soup is clear and transparent. Based on a typical French method, and in line with the unique clear soup, delicious, light principles of soup materials can be made into different varieties.

❷ Puree soup

Boil vegetables and beans in clear water, grind and strain them, and then add in clear soup or cream soup and salted pork skin over low heat. When served, garnish the soup with fried bread crumbs or frothy cream.

❸ Vegetable soup

A soup made from meat and vegetables followed by clear soup. This kind of soup mostly carries some meat, also known as meat vegetable soup.

❹ Cream soup

A cream soup is made from oily fried flour with milk and some condiments, based on which fish, chicken and vegetable puree are added to make a variety of cream soup.

⑤ Thick soup (broth)

Broth is a savory liquid made of water in which bones, meat, fish or vegetables have been simmered. It can be eaten alone, but is most commonly used to prepare other dishes such as the gravy and sauce.

⑥ Seafood soup/Fish soup

Fish soup is a warm food made by combining fish or seafood with vegetables and stock, juice, water, or other liquid.

⑦ Iced soup

Iced soup is made from clear soup or cold boiled water, added to vegetables and a little cooked meat and then mixed, such as cold fresh tomato soup, which is often seen in Russian restaurants.

烹饪英语

Exercises

Activity Translate the phrases into English.

1. 清汤菜丝 _____
2. 法式洋葱汤 _____
3. 甘笋茸汤 _____
4. 牛尾汤 _____
5. 鸡杂浓汤 _____
6. 蔬菜忌廉汤 _____
7. 鲜番茄冷汤 _____
8. 洋葱汤 _____
9. 鸡肉蔬菜汤 _____
10. 龙虾浓汤 _____

Part B French Onion Soup

Learning Goals

You will be able to:

1. know the soup's name;
2. learn how to make french onion soup.

Vocabulary Assistance

onion [ˈʌnjən]   n. 洋葱（头）　　　　broth [brɒθ]   n. 肉汤
butter [ˈbʌtə(r)]   n. 黄油　　　　　　toast [təʊst]   v. 烘，爆
pepper [ˈpepə(r)]   n. 胡椒粉,辣椒　　cheese [tʃiːz]   n. 奶酪
bay [beɪ]   n. 月桂　　　　　　　　　thyme [taɪm]   n.（用以调味的）百里香

simmer ['sɪmə(r)] n. & v. 炖,慢煮
sprinkle ['sprɪŋkl] v. 洒
reduce [rɪ'djuːs] v. 减少,降低
meantime ['miːntaɪm] n. 同时,其间
cover ['kʌvə(r)] v. 覆盖,代替
fascinating ['fæsɪneɪtɪŋ] adj. 吸引人的
add [æd] v. 加,加入,增加,添加
occasionally [ə'keɪʒnəli] adv. 不时,有时

Dialogue

A: What are you doing?
B: I'm making soup.
A: What soup?
B: French onion soup.
A: The soup sounds rather fascinating. How do you prepare it?
B: First slice 40 onions. Then cook the onions in butter.
A: Over low heat?
B: Yes, and stir occasionally.
A: For how long?
B: 20 or 30 minutes.
A: What's next?
B: Add the beef broth, water, bay leaves, pepper and thyme. Then heat it to boil. Reduce the heat. Cover and simmer. In the meantime, you can toast the French bread.
A: How long?
B: About 15 minutes. Then put the toasted French bread into bowls and pour onion soup on top.
A: Is that all?
B: No. The last step is to sprinkle Swiss and Parmesan cheeses on top and put the bowls in the salamander just before serving.
A: Oh, I've got it. Thank you!
B: Not at all.

Exercises

Activity Write the recipe and cooking procedures according to the dialogue.

French Onion Soup

1. 原材料：

_____
_____

2. 制作过程：

_____
_____
_____

Part C Chicken Soup

 Learning Goals

You will be able to：

1. know the soup's name；
2. learn how to make chicken soup.

 Vocabulary Assistance

lemon ['lemən]   n. 柠檬
juice [dʒuːs]   n. 汁，果汁
prepare [prɪ'peə(r)]   v. 准备
saucepan ['sɔːspən]   n. 炖锅
delicious [dɪ'lɪʃəs]   adj. 美味的，可口的
break [breɪk]   v. 打破，打断

hold [həʊld]   v. 拿着，握住
stir [stɜː(r)]   v. 搅拌
immediately [ɪ'miːdɪətli]   adv. 立即，马上
boned [bəʊnd]   adj. 去骨的
red date   红枣
ingredient [ɪn'griːdɪənt]   n. 成分，原料

 Dialogue

A：Hi, Chef. Can you make chicken soup?

B：Of course, you know I'm a good cook.

A：Wow, please teach me, thanks.

B：OK. First, you need to prepare ingredients: boned chicken, butter, onion, red date, pepper, lemon juice, egg and salt. Now, let's start. Put the chicken into the saucepan with butter, salt, red date, onion and a bit of pepper. Cook it for 8 minutes. Add water, lentils and lemon juice and boil it for 1.5 hours. Beat an egg and pour it into the soup. The delicious soup is done then.

A：Perfect. Thank you very much. It's so kind of you.

B：You are welcome.

Exercises

Activity Find a partner to ask and answer the following questions according to the dialogue.
1. What will we make?
2. What are the ingredients?
3. Can you tell me what to do first?
4. How much pepper do they need?
5. Do you know how to make chicken soup now?
6. What kind of soup do you often make?

Part D Eel Soup

Learning Goals

You will be able to:
1. know the soup's name;
2. learn how to make eel soup.

Vocabulary Assistance

eel [iːl]   n. 鳗鱼
bacon ['beɪkən]   n. 咸猪肉,熏肉
conger eel   大海鳗
prune [pruːn]   n. 梅脯,梅干
skin [skɪn]   v. 去皮,剥去……的皮
sprig [sprɪɡ]   n. (烹饪或装饰用)小枝
chunk [tʃʌŋk]   n. (厚)块
pear [peə(r)]   n. 梨

dried [draɪd]   adj. 弄干了的
smoke [sməʊk]   n. & v. 烟,熏
ham [hæm]   n. 火腿
leek [liːk]   n. 韭葱
white wine   白葡萄酒
plain [pleɪn]   adj. 普通的,不掺杂的
flour ['flaʊə(r)]   n. 面粉
chervil ['tʃɜːvɪl]   n. 细叶芹属植物,雪维菜,细叶片

marjoram ['mɑːdʒərəm] n. 墨角兰,牛至	work [wɜːk] v. 揉(面团),搅(黄油)
savory ['seɪvərɪ] n. 香薄荷	smooth [smuːð] adj. 光滑的,调匀的
tarragon ['tærəgən] n. 龙蒿	paste [peɪst] n. (做点心等用的)加了油脂的面团
vinegar ['vɪnɪgə(r)] n. 醋	
partly ['pɑːtli] adv. 部分地,不完全地	tiny ['taɪni] adj. 极小的
continue [kən'tɪnjuː] v. 继续,连续	piece [piːs] n. 碎片
gently ['dʒentli] adv. 柔和地,轻轻地	slightly ['slaɪtli] adv. 轻微地
soft [sɒft] adj. 软的	thicken ['θɪkən] v. 使变厚(或粗、密、浓)
rub [rʌb] v. 擦,搓	liquor ['lɪkə(r)] n. 液体,汁,烈性酒
tender ['tendə(r)] adj. 嫩的,柔软的	seasoning ['siːzənɪŋ] n. 调味品
meat [miːt] n. 食用肉类,(猪、牛、羊等的)鲜肉	wine [waɪn] n. 葡萄酒,深红色
	dry [draɪ] adj. 无甜味的
aside [ə'saɪd] adv. 在旁边	sugar ['ʃʊgə(r)] n. 糖

Ingredients

1 smoked ham bone or half a pound of bacon

8 ounces of soaked prunes

8 ounces of soaked dried pears

1 thinly sliced large onion

4 thinly sliced carrots

1 thinly sliced leek

8 ounces of shelled green peas

2½ pounds of eel

1 bay leaf

4 black peppercorns

5 fluid ounces of dry white wine

1 tablespoon of butter

1 tablespoon of plain flour

3 finely chopped sprigs of parsley

2 finely chopped sprigs of chervil

1 finely chopped sprig of marjoram

1 finely chopped sprig of thyme

1 finely chopped sprig of savory

1 finely chopped sprig of tarragon

1 finely chopped sprig of basil

sugar

salt

wine vinegar

Text

In a large saucepan, boil the ham bone or bacon in 7 pints of water, partly covered, for 2 hours. Add the fruits and vegetables, and continue to simmer gently until they are soft.

In the meantime, skin the eel and rub it with salt, cut it into chunks and place it in another pan with the bay leaf, peppercorns, wine and enough water to cover it completely.

Place the pan over high heat and bring it to a boil. Cover the pan, reduce the heat to low and simmer gently until the eel is tender.

Remove the ham bone or bacon from the soup. Cut the meat into chunks and set it aside. Work the butter and flour into a smooth paste. Add it to the soup in tiny pieces, stirring until the soup is slightly thickened. Stir in the finely chopped herbs.

Add the cooked eel together with its cooking liquor (juice) and the meat from the ham bone or the bacon.

Taste the soup and correct the seasoning. Add more salt if necessary. Also add a pinch of sugar and a few drops of wine vinegar.

Exercises

Activity Ask and answer the following questions with a partner, using a complete sentence.

1. What kind of pears do they need?
2. How many green peas do they want?
3. How long will they boil the ham?
4. When will they add the fruits and vegetables?
5. Will they skin the eel?
6. How do they cook the eel?
7. Why do they add the cooked eel together with its juice?
8. Have you ever had eel soup?
9. Who will taste the soup first, the chef or the commis?

Unit 9

Pastry

Part A Chinese-style Pastry

 Learning Goals

You will be able to:
1. know the classification of Chinese-style pastry;
2. tell the name of the Chinese-style pastry;
3. know how to cook Chinese-style pastry.

 Vocabulary Assistance

pastry ['peɪstri] n. 油酥点心	a layer of 一层
knead [niːd] v. 揉,捏	salt [sɔːlt] n. 盐
dough [dəʊ] n. 生面团,面浆团	cylinder ['sɪlɪndə(r)] n. 圆柱体
damp cloth 湿布	lard [lɑːd] n. 猪油
elastic dough 水调面团	fry [fraɪ] v. 油煎
rinse [rɪns] n. 冲洗	cooking oil 食用油
puffy dough 膨松面团	dumpling ['dʌmplɪŋ] n. 饺子
mince [mɪns] v. 剁碎	ingredient [ɪnˈɡriːdiənt] n. 成分,原料
pastry dough 油酥面团	shrimp [ʃrɪmp] n. 虾
strip [strɪp] n. 细长条	wonton [ˌwɒnˈtɒn] n. 馄饨
rice flour dough 米粉面团	hemp flower 麻花
equal portion 等份	walnut cake 桃酥
scallion ['skæliən] n. 葱	walnut ['wɔːlnʌt] n. 核桃
flat [flæt] adj. 平的	fried dough sticks 油条
flour ['flaʊə(r)] n. 面粉	crispy durian pastry 榴莲酥
grease [ɡriːs] v. 抹油	durian ['dʊəriən] n. 榴莲
all-purpose 通用的	

Text

How to make scallion cake

Ingredients:

all-purpose flour
boiling water
cold water
salt
scallion
lard or cooking oil

Methods:

Step 1. Combine all-purpose flour, boiling water, cold water and salt well, and knead into smooth dough. Cover it with a damp cloth and let it sit for 20 minutes.

Step 2. Rinse scallions well and mince them. Knead the dough more smoothly and roll out into long strips. Then divide the strips into 5 equal portions. Roll each portion out wide and flat and grease with a layer of lard (or cooking oil) evenly, then spread with a layer of minced scallion. Roll up into a strip. Then roll up from both sides into a cylinder.

Step 3. Press the stuffed dough flat and roll into a flat cake. Fry in frying pan with a little oil added until it gets golden on both sides. Remove and cut into sections.

Dialogue

Waiter: Good evening, sir. May I take your order?

Guest: Yes, what do you recommend?

Waiter: Our traditional dumpling is popular. They're steamed stuffed buns filled with pork and shrimp meat.

Guest: Sounds great! I'll have it.

Waiter: Would you like some soup?

Guest: Yes. But I prefer salty soup.

Waiter: Egg soup or the lamb soup would be the best. Which would you prefer?

Guest: Egg soup is OK. Thanks.

Waiter: Sure. I will serve immediately.

Exercises

Activity I What Chinese-style pastries can you see in the pictures? Please write them down, then translate them into English.

1. _____ 2. _____ 3. _____

4. _____ 5. _____ 6. _____

7. _____ 8. _____ 9. _____

Activity Ⅱ   Classify these Chinese-style pastries in the above pictures.

Elastic dough: _____

Puffy dough: _____

Pastry dough: _____

Activity Ⅲ

Fried dough sticks

　　Fried dough sticks are elongated, yellow and hollow. It smells delicious. It tastes very crispy and very well. People usually eat with soybean milk. In China, KFC and McDonald's sell fried dough sticks. Usually, people can also buy them in some breakfast bars.

【Vocabulary】
fried dough sticks 油条
elongate ['i:lɒŋgeɪt] v.(使)变长,伸长,拉长
hollow ['hɒləʊ] adj. 中空的,空心的
soybean milk 豆浆

【Questions】
1. What are the characteristics of fried dough sticks?
_____
2. Where can buy fried dough sticks?
_____

Focus on Language

Listen to the following passage and fill in the blanks.

The jiaozi is a common Chinese 1._____ which generally consists of minced meat and chopped 2._____ wrapped into a piece of 3._____. Popular meat fillings include ground 4._____, ground 5._____, ground chicken, 6._____, and even fish. Popular mixtures include pork with Chinese 7._____, pork with celery, lamb with spring onion, leeks with 8._____. Dumplings are usually boiled or steamed. Dumpling is a 9._____ dish eaten during Chinese New Year's Eve and some other 10._____. Family members gather together to make dumplings. It is also eaten for farewell to family members or friends.

Part B Western-style Pastry

Learning Goals

You will be able to:
1. know the classification of Western-style pastry;
2. tell the name of the Western-style pastry;
3. know how to cook Western-style pastry.

Vocabulary Assistance

a cup of 一杯
alternate [ɔːlˈtɜːnət] v. 交替
cocoa powder 可可粉
bake [beɪk] v. 烘烤
preheat [ˌpriːˈhiːt] v. 预热
toothpick [ˈtuːθpɪk] n. 牙签
oven [ˈʌvn] n. 烤箱

frost [frɒst] v. 撒上糖霜
rack [ræk] n. 支架
tiramisu [ˌtɪrəˈmiːsuː] n. 提拉米苏
baking soda 小苏打
pudding [ˈpʊdɪŋ] n. 布丁
flour [ˈflaʊə(r)] n. 面粉
pizza [ˈpiːtsə] n. 比萨

外教有声

butter [ˈbʌtə(r)]   n. 黄油
seafood [ˈsiːfuːd]   n. 海鲜
mixer [ˈmɪksə(r)]   n. 搅拌器
latte [ˈlɑːteɪ]   n. 拿铁
vanilla [vəˈnɪlə]   n. 香草香精
mocha [ˈmɒkə]   n. 摩卡咖啡
paddle [ˈpædl]   n. 桨叶
attachment [əˈtætʃmənt]   n. 附件
bowl [bəʊl]   n. 碗
cheese [tʃiːz]   n. 奶酪
incorporate [ɪnˈkɔːpəreɪt]   v. 合并

medium rare   三分熟
sour cream   酸奶油
baguette [bæˈget]   n. 法棍
egg tart   蛋挞
pasta [ˈpæstə]   n. 意大利面
waffle [ˈwɒfl]   n. 华夫饼
brioche [briˈɒʃ]   n. 布里欧修
croissant [krwɑːˈsɑː]   n. 牛角面包
specialty [ˈspeʃəlti]   n. 特色，专长
mozzarella [ˌmɒtsəˈrelə]   n. 莫泽雷勒干奶酪
cappuccino [ˌkæpuˈtʃiːnəʊ]   n. 卡布奇诺

 Text

How to make the birthday cake

Ingredients：

cocoa powder
boiling water
all-purpose flour
baking soda
salt
butter
sugar
vanilla
egg
sour cream

Methods：

Step 1. Preheat your oven to 350 ℃ and set a rack in the middle of the oven.

Step 2. Put butter and flour in round pans and whisk.

Step 3. In another bowl, whisk together the flour, baking soda and salt.

Step 4. Use your stand mixer with the paddle attachment and the speed set to medium, beat the butter, sugar and vanilla until light and fluffy.

Step 5. Add one egg at one time, waiting until the first egg is fully incorporated before adding the second one.

Step 6. Reduce the mixer speed to low and add in the flour and the sour cream in alternating thirds.

Step 7. Add the cocoa powder mixture.

Step 8. Pour half of the butter into the pans.

Step 9. Bake the cakes for 40 minutes or until a toothpick inserted into the center of the cake comes out clean.

Step 10. Cool the cakes for 20 minutes.

Step 11. Remove the cakes from the pans and let cool completely on a rack.

Step 12. Frost the cakes with your favorite frosting, creating a two-layer cake.

Dialogue

Waiter: Good evening, madam. May I take your order?

Guest: Yes, what is your special cuisine today?

Waiter: The special cuisine today is our regular set meal. It includes steak, pizza, cake and coffee.

Guest: All right. I'll have it.

Waiter: How would you like your steak?

Guest: I'll have it medium rare please.

Waiter: Would you prefer fruit pizza or seafood pizza?

Guest: Fruit pizza.

Waiter: Would you prefer tiramisu or pudding?

Guest: Tiramisu, please.

Waiter: What kind of coffee would you like, madam? We have latte, mocha and cappuccino.

Guest: Cappuccino, thanks.

Waiter: It's my pleasure.

Exercises

Activity I What Western-style pastries can you see in the pictures? Please write them down, then translate them into English.

1. _____

2. _____

3. _____

4. _____

5. _____

6. _____

7. _____   8. _____   9. _____

Activity Ⅱ

Egg tart

The egg tart is regarded as one of Britain's traditional dishes. It remained a favourite over the centuries and is just as popular today. Egg tarts are usually made from shortcrust pastry, eggs, sugar, milk or cream, and vanilla, sprinkled with nutmeg and baked. To make them a special treat they can be topped with fresh strawberries or raspberries. The egg tart is a lovely treat to come home to and goes wonderfully with a cup of Earl Grey tea. You just can't get more English than that.

【Vocabulary】

shortcrust pastry　酥皮糕点
shortcrust　adj. 酥脆的
sprinkle [ˈsprɪŋkl]　v. 撒，洒
nutmeg [ˈnʌtmeg]　n. 肉豆蔻
raspberry [ˈrɑːzbəri]　n. 覆盆子
Earl Grey tea [ˌɜːrlˈgreɪ tiː]　伯爵茶

【Questions】

1. What materials are usually used to make egg tarts?

_____

2. What kind of drink should egg tarts go with?

_____

Focus on Language

🎧 **Listen to the following passage and fill in the blanks.**

Italian pizza may be widely known to have invented only the last 1._____ of the 19th century, although some believe that it was the 2._____ who first made them, when a cook was summoned by his King and Queen to make a local 3._____. He then made rounded bread with topping and colored it with Italian flag colors. He used Italian ingredients to make the colors particularly the 4._____ sauce for red, mozzarella 5._____ for white and basil 6.____ for green. The 7._____ was named after the Queen Mergherita and was 8._____ as the first Italian pizza. From that time on, 9._____ pizza has become the most 10._____ pizza around the world.

Part C Mounting Decoration

Learning Goals

You will be able to:
1. know the classification of mounting decoration cake;
2. tell the name of the tool of mounting decoration;
3. know how to cook mounting decoration cake.

Vocabulary Assistance

mount [maʊnt]　v. 安装
spatula [ˈspætʃələ]　n. 抹刀
decoration [ˌdekəˈreɪʃn]　n. 装饰
slice [slaɪs]　v. 切片
piping bag [ˈpaɪpɪŋ bæg]　裱花袋
grain [greɪn]　n. 纹理
Chiffon cake　戚风蛋糕
nozzle [ˈnɒzl]　n. 裱花嘴
light cream　淡奶油
cherry [ˈtʃeri]　n. 樱桃
strawberry [ˈstrɔːbəri]　n. 草莓
drain [dreɪn]　v. 沥干
cake plug-in　蛋糕插件
cake base [beɪs]　蛋糕坯
hydration [haɪˈdreɪʃn]　n. 水化
8-tooth [tuːθ]　8齿

mounting turntable [ˈtɜːnteɪbl]　裱花转台
squeeze [skwiːz]　v. 挤压
remaining [rɪˈmeɪnɪŋ]　adj. 剩下的
plaster [ˈplɑːstə(r)]　v. 抹面
appropriate [əˈprəʊpriət]　adj. 合适的
wipe [waɪp]　v. 抹
rotate [rəʊˈteɪt]　v. (使)旋转
erect [ɪˈrekt]　v. 竖起
pitaya [ˈpɪtəjə]　n. 火龙果
pull wire　拉线
Korean decoration　韩式裱花
toothpick [ˈtuːθpɪk]　n. 牙签
cream decoration　奶油裱花
dip [dɪp]　v. 蘸
chocolate decoration　巧克力裱花
shade [ʃeɪd]　n. 色度

duplicate [ˈdjuːplɪkeɪt]　v. 复制
swirl [swɜːl]　v. 旋转
intensify [ɪnˈtensɪfaɪ]　v.（使）加强
blend [blend]　v. 混合

canned yellow peach　黄桃罐头
horizontally [ˌhɒrɪˈzɒntəli]　adv. 水平地，横向地
electric eggbeater　电动打蛋器

Text

How to make a mounting decoration cake

Ingredients：
Chiffon cake
light cream
sugar
strawberry
cherry
cake plug-in
canned yellow peach

Methods：

Step 1. Drain the canned yellow peach and set it aside. Too much water will cause the cream to melt easily and affect the taste of Chiffon cake.

Step 2. Cut Chiffon cake into 3 pieces horizontally.

Step 3. Add sugar into the cream, and beat it with electric eggbeater at medium speed until the grain is clear. If the cream is beaten, it cannot be used because of hydration.

Step 4. Put cake base on the mounting turntable, and apply a layer of cream on the first layer of cake base with a spatula.

Step 5. Spread the canned yellow peach evenly, or use other fruits instead. Put a second layer of cake on top. In the same way, spread the cream on the second layer of cake and arrange the fruits. Set the top cake and the sandwich is finished.

Step 6. For plastering, firstly wipe the four sides and the top, and then apply appropriate amount of cream around. Secondly, erect the spatula, rotate the mounting turntable, and try to smooth it as much as possible. Thirdly, put the appropriate amount of cream on the top, hold the spatula in hand, and keep it level as far as possible. Rotate the mounting turntable and carefully arrange the edges and corners.

Step 7. Slice the fresh strawberries and place them around the bottom of the cake. Put the remaining cream into the mounting bag, choose the medium 8-tooth mounting mouth, and squeeze a circle on the outer edge of the cake. Then place fresh strawberries in the middle. Place cherries around and insert the cake plug-in. The three-layer sandwich fruit cake is ready.

Dialogue

Waiter: Good afternoon, madam. What kind of cake would you like to order?

Guest: Good afternoon, what do you recommend?

Waiter: We have chocolate decoration cake, cream cake, sugar noodle cake, finger cream cake. Which flavor do you prefer?

Guest: I prefer chocolate flavor.

Waiter: Sure. Tell us what you want to write on it, and then we'll see if it's possible.

Guest: "Happy birthday to you" is OK.

Waiter: What fruits are in the cake?

Guest: What kind of fruit do you have?

Waiter: We have yellow peach, mango, pitaya and strawberry.

Guest: I would like pitaya and strawberry.

Waiter: Sure.

Guest: When will it be ready?

Waiter: It will be ready about half an hour later.

Guest: Oh, that's fine. I'll be back for it half an hour later.

Waiter: OK. Then the cake will be ready. See you later.

Guest: See you later.

 Exercises

Activity Ⅰ What tools and types of decoration can you see in the pictures? Please write them down, then translate them into English.

1. _____   2. _____   3. _____

4. _____   5. _____   6. _____

7. _____ 8. _____ 9. _____

Activity Ⅱ Classify these tools and types of decoration in the above pictures.

Tools: _____

Types of decoration: _____

Focus on Language

Listen to the following passage and fill in the blanks.

Basic color mixing

Begin with white 1._____ and add 2._____ a little at a time until you achieve the 3._____ you desire. Use a 4._____ to add cream color. Just 5._____ it into the color, swirl it in the cream, and blend well with 6._____. Always use a new toothpick when adding more colors. Always mix enough of the color cream you will need to complete the project. It's difficult to duplicate an exact shade of any color. To 7._____ deep or dark color, you will need to add color in large 8._____. Colors will 9._____ or darken in buttercream, so keep this in 10._____ when mixing colors.

Unit 10

Menu Recommendation

扫码看课件

Part A Starters

 Learning Goals

You will be able to:
1. recommend common starters;
2. explain the main ingredients of common starters;
3. know how to translate the dish names of starters.

 Vocabulary Assistance

cocktail ['kɒkteɪl] n. 鸡尾酒,开胃食品
appetizer ['æpɪtaɪzə(r)] n. 吊胃口的东西,开胃食品
hors d'oeuvre [ɔː'dɜːv] n. (法)(主菜前的)开胃小菜
entremets [ˌɒntrə'meɪ] n. 两道菜之间上的一道清淡的菜,甜点

savory ['seɪvəri] n. (聚会上提供的)咸味小吃
precooked [ˌpriː'kʊkt] adj. (食物)煮好的
reheated [ˌriː'hiːt] v. 将……重新加热
buffet ['bʊfeɪ] n. 自助餐

 Text

    A starter, appetizer, or hors d'oeuvre is a small dish served before a meal in European cuisine. Some starters are served cold, others hot. Starters may be served at the dinner table as a part of the meal, or they may be served before seating, such as at a reception or cocktail party. Formerly, starters were also served between courses. Typically smaller than a main dish, a starter is often designed to be eaten by hand.

    During the Middle Ages, formal French meals were served with entremets between the servings of plates. These secondary dishes could be either actual food dishes, or elaborate displays and even dramatic or musical presentations. The custom of the savory course is of British origin and comes towards the end of the meal, before dessert or sweets or even after the dessert, in contrast to the hors d'oeuvre, which is served before the meal.

    In restaurants or large estates, starters are prepared in a grade manger which is a cool room. Starters are often prepared in advance. Some types may be refrigerated or frozen and then precook and reheat in an oven or microwave oven as necessary before serving.

It is also an unwritten rule that the dishes served as starters do not give any clue to the main meal. They are served with the main meal menu in view either in hot, room temperature or cold forms; when served hot they are brought out after all the guests arrive so that everyone gets to taste the dishes.

Dialogue

Waiter: Good evening, sir. May I ask you a total number of people to dine?
Guest: Just two, do you have a table?
Waiter: Over there, sir. Please be seated.
Guest: Thank you. We are not so hungry. Can you recommend something for me?
Waiter: What about Roman style chicken? It's our new dish.
Guest: Sounds great! We could have a try.
Waiter: What would you like for appetizer and drink?
Guest: Shrimp cocktail for appetizer and red wine for drink, please.
Waiter: OK. Please wait for a while. We'll get it done quickly!

Exercises

Activity I Fill in the blanks according to the pictures and clues given.

1._____ in aspic   2.Caesar _____   3._____ salad

4.stuffed _____   5.pickled _____   6.____ and ____ spare ribs

7.mixed green _____ 8.cucumber _____ 9.marinated _____

10.grilled spicy _____ with spinach tomato salad   11.baked _____ with cheese and white sauce   12.spice roast _____

Activity II

Role playing A

W: Would you like some appetizers before your meal?

C: I'd like to try some _____. (black fungus/cucumber in sauce/mixed salad)

Role playing B

C: Would you please offer some _____ for us before meal. (cold dishes/appetizers)

W: How about _____? (salad/fruit cocktail/smoked salmon)

C: It sounds great, I'll have one.

Role playing C

C: I just want to have a _____ meal here. What do you recommend? (light/spicy/sweet)

W: How about some _____? (dip/pickle/aspic)

Activity III Translate cocktail party menu.

1. 干果：烤杏仁、炸花生米、炸桃仁、炸土豆片

2. 冷小吃：火腿、冷肉肠、熏三文鱼、鸡肝派、酸黄瓜

3. 热小吃：炸大虾、小汉堡

English	Chinese	English	Chinese
foie gras in aspic	鹅肝冻	mixed green beans	凉拌毛豆
Caesar Salad	凯撒沙拉	cucumber rolls	黄瓜卷

续表

English	Chinese	English	Chinese
combination salad	什锦沙拉	marinated tripe	卤水毛肚
pickled cucumbers	凉拌酸黄瓜	grilled spicy prawns with spinach tomato salad	铁扒大虾配番茄沙拉
stuffed tomato	填馅番茄	baked oysters with cheese and white sauce	芝士焗生蚝配白汁
sweet and sour spare ribs	糖醋排骨	spice roast beef	五香牛肉

Part B Main Course

 Learning Goals

You will be able to:
1. recommend common main course;
2. explain the main ingredients of common main course;
3. know how to translate the dish names of main course.

 Vocabulary Assistance

primary ['praɪməri]   adj. 主要的
course [kɔːs]   n. 一道菜
meatloaf ['miːtləʊf]   n. 烘肉卷

substantial [səb'stænʃl]   adj. 大量的
gastronomic [ˌɡæstrə'nɒmɪk]   adj. 烹饪的

 Text

The main course is the featured or primary dish in a meal consisting of several courses. It usually follows the entrée ("entry") course. Americans think of an entrée as the main course — the meatloaf or the roast chicken. But the French word actually means "entrance". On a menu in France, an entrée is more of an appetizer. The original meaning of entrée — as it was used during the late middle ages and early Renaissance — was much closer to the American meaning.

The main dish is usually the heaviest, heartiest, and most complex or substantial dish in a meal. The main ingredient is usually meat, fish or another protein source. It is most often preceded by an appetizer, soup or salad, and followed by a dessert. For those reasons the main course is sometimes referred to as the "meat course". In formal dining, a well-planned main course can function as a sort of gastronomic apex or climax. In such a scheme, the preceding courses are designed to prepare for and lead up to the main course in such a way that the main course is anticipated, when the scheme is successful, increased in its ability to satisfy and delight the diner. The courses following the main course then calm the palate and the stomach, acting as a sort of denouement or anticlimax.

Dialogue

In hot kitchen

(A-Mr. Jone B-Chef Mr. Wang C-Chef Lao Chan D-Chef Xiao Zhao)

A: Hi, here is a menu, a beef steak with black pepper sauce, a rare one, please.

B: Yes, it will be ready soon. Please help me garnish some vegetables.

D: OK. The vegetables are ready, serve it now, please.

A: Lao Chan, please cook a baked salmon, more cheese, please.

C: I see, please wait 10 minutes. Bring a baking pan for me please.

D: The stainless or the China pan?

C: China pan, please. Don't forget to spread some butter on it.

A: Mr. Wang, please put some cheese into the black pepper sauce but not too much. Put more black pepper please.

B: I see.

D: Hi, Mr. Jone, may I serve the soup now?

A: OK. What soup?

D: Crab-meat potage with cream. Shall I garnish it?

A: Yes, please garnish it with parsley.

Exercises

Activity I Fill in the blanks according to the pictures.

1. grilled sirloin _____ with vegetables

2. _____ ball

3. grilled _____

4. fried _____ chop

5. _____ Wellington

6. grilled prawns with ____

7.grilled young _____   8._____ platter   9._____ mandarin fish

10.stewed ___ with shrimps   11.roast _____   12.roast suckling_____

Activity Ⅱ Fill in the blanks with conjunctions.
1. 奶油烩香肠　stewed sausage _____ cream
2. 火腿煎蛋　fried eggs _____ ham
3. 萝卜排骨汤　pork ribs _____ turnip soup
4. 竹笋青豆　bamboo shoots _____ green beans
5. 罐焖山鸭　pheasant _____ casserole
6. 红酒腌肉蜗牛　snails _____ bacon _____ red wine sauce
7. 白汁烩肉　meat stewed _____ white sauce

Tips:菜名翻译为主料＋连词(and/with)＋辅料(＋连词(in/with)＋汤汁/器具)

English	Chinese	English	Chinese
grilled sirloin steak with vegetables	铁扒西冷牛排配时蔬	grilled young pigeon	铁扒乳鸽
shrimps ball	炸虾球	sashimi platter	刺身拼盘
grilled lobster	铁扒大虾	steamed mandarin fish	清蒸鳜鱼
fried chicken chop	炸鸡肉卷	stewed sea cucumbers with shrimps	虾仁扒海参
beef Wellington	惠灵顿牛肉	roast duck	烤鸭
grilled prawns with asparagus	铁扒大虾配芦笋	roast suckling pig	烤乳猪

常见时令菜单

Cold dish	
Vegetable salad	蔬菜沙拉
Country salad	乡村沙拉
Prawn salad	大虾沙拉
Scallops and tender spinach salad	嫩菠菜鲜贝沙拉
Foie gras in aspic	鹅肝冻
Ham with egg	鸡蛋火腿
Salmon	三文鱼
Terrine of Foir Gras	鹅肝酱
Soup	
Ox-tail soup	牛尾汤
Cream of vegetable soup	意大利浓汤
Mashed spinach soup	奶油菠菜汤
Fish	
Grilled sole	铁扒比目鱼
Prawns with saffron	藏红花大虾
Baked of prawns	焗明虾
Scallops Provencal style	普罗旺斯式鲜贝
Fried turbot with butter	黄油煎菱鲆鱼
Meet	
Lamb cutlet with thyme	百里香牛排
Grilled beef fillet	铁扒菲力牛排
Green pepper steak	青椒牛排
Rib steak Bercy style	贝西牛排
Roast duck with orange	橙子烤鸭
Vegetable	
Green beans in butter	黄油扁豆
Boiled peas French style	法式煮豌豆
Mashed vegetable of the day	时令蔬菜泥
Fried potato strips	炸土豆条

常见节日菜单

感恩节套餐 Thanksgiving Day set menu
Apple & pumpkin soup with pine nuts cream
苹果南瓜汤
Scallop crepe with pesto
扇贝配罗勒酱
Toasted turkey & smoked duck breast with gravy
烤火鸡烟熏鸭胸伴传统烧汁
Thanksgiving carrot cake
感恩胡萝卜饼
Petit fours
精美小甜品
Coffee or tea
咖啡或茶

情人节菜单 Valentine's Day set menu
Foie gras terrine with truffle
松露菌鹅肝酱
Baked turtle & black garlic soup with puff pastry
酥皮焗黑金水鱼汤
Grilled U.S beef tenderloin with king prawn
扒美国牛柳拼珍宝虾
Strawberry "heart" cheese cake
草莓爱心芝士饼
Petit fours
精美小甜点
Coffee or tea
咖啡或茶

Answers

Unit 1 Introduction to Cooking Industry

Part A

Activity Ⅰ 略

Activity Ⅱ

1. indispensable 2. development 3. component 4. represent 5. prestigious 6. revolution 7. civilization 8. disease

Activity Ⅲ 略

Part B 略

Unit 2 Kitchen Introduction

Part A

Activity Ⅰ 略

Activity Ⅱ

1. Relief cook 2. Executive chef 3. Larder chef 4. Pantry man 5. Butcher 6. Breakfast cook 7. Potman 8. Roast cook 9. Pastry chef 10. Commis

Focus on Language

1. Executive chef 2. Sous chef 3. Sommelier 4. Personal chef 5. Master chef 6. Pastry chef 7. Prep cook 8. Catering chef 9. Baker 10. Fry cook 11. Sauce chef

Part B

Activity Ⅰ 略

Activity Ⅱ

1. wet 2. soap 3. lather 4. scrub 5. rinse 6. dry

Focus on Language

1. be aware of routinely cleaning it 2. hygienic environment
3. clean as you go 4. cross-contaminate
5. use a different chopping board 6. hot water and dish washing liquid
7. rinse in clean, hot water

Part C

Activity Ⅰ

1. safe 2. safe 3. danger 4. danger 5. safe

Activity Ⅱ

safe 2、3、5、9

unsafe 1、4、6、7、8、10

Focus on Language

1. dangerous　2. sharp instruments　3. traffic　4. containers
5. items　6. a couple of days　7. Spills

Unit 3　Tools and Equipment

Part A

Activity Ⅰ

1. colander　2. meat hammer　3. roasting fork　4. tongs　5. sieve
6. bench scraper or dough knife　7. corkscrew　8. mixing bowl　9. frying basket　10. ladle
11. whisk　12. spatula

Activity Ⅱ

A	B
rolling pin	cut meat or vegetables
meat hammer	cut pieces of dough
chopping board	beat meat
frying basket	make French fries
bench scraper	make dough flat

rolling pin — make dough flat
meat hammer — beat meat
chopping board — cut meat or vegetables
frying basket — make French fries
bench scraper — cut pieces of dough

Part B

Activity Ⅰ

Crepe Recipe 法式薄饼

Ingredients：原料

4 eggs　4个鸡蛋

1 cup of flour　1杯面粉

1/2 cup of milk　半杯牛奶

1/2 cup of water　半杯水

1/2 teaspoon of salt　半茶匙盐

2 tablespoons of melted butter　两汤匙融化的黄油

Preparation：准备工作

1. Measure all ingredients into blender jar, blend for 30 seconds.

称取所有的原料后放入搅拌容器内，搅打30秒。

2. Scrape down sides. Blend for 15 seconds more. Cover and let sit for 1 hour (this helps the flour absorb more of the liquid.)

将搅拌容器内壁的残留混合物刮下来后，再搅拌15秒。盖上盖子后静置1小时（帮助面粉吸收更多的液体）。

Regular Crepe Pan Directions：常规说明

1. Heat crepe pan and grease lightly.

预热平底锅，涂少许润滑脂。

2. Measure about 1/4 cup of batter into pan. Tilt pan to spread batter. Once crepe has lots of little bubbles, loosen any edges with the spatula. Flip crepe over. This side cooks quickly. Slide crepe from pan to plate.

把大约 1/4 杯的面糊倒入锅里。倾斜平底锅铺开面糊。一旦有很多小气泡,就用抹刀把边缘弄松。把薄饼翻个面,这面煎得很快。将薄饼从平底锅滑到盘里。

Smoked Salmon Chowder 烟熏三文鱼杂烩

Ingredients:原料

a knob of butter　一小块黄油

1 onion, finely diced　1 个洋葱,切碎的

750 grams of potatoes, diced　750 克土豆,切成丁状

500 milliliters of chicken stock　500 毫升基础鸡汤

500 milliliters of milk　500 毫升牛奶

350 grams of smoked salmon, cut into ribbons　350 克烟熏三文鱼,切成条状

a handful of parsley leaves, finely chopped　一把欧芹叶,切碎的

1 lemon, halved　1 个柠檬,分成两半

Methods:制作方法

1. Fry the onion gently with butter in the sauté pan, then add potatoes, stock and milk, and simmer until potatoes are very tender.

在炒锅里炒香洋葱碎,然后加入土豆、基础鸡汤、牛奶一起炖直到土豆变软。

2. Add the smoked salmon and parsley leaves, and season well.

加入烟熏三文鱼、欧芹叶后调味。

3. Heat everything through and add a squeeze of lemon.

热透所有的原料,最后挤上柠檬汁。

Activity Ⅱ

1. sauce pan　2. sauté pan　3. frying pan　4. brazier　5. stockpot

Part C

Activity

1. butter knife　2. Chinese cook's knife　3. vegetable peeler

4. chef's knife　5. paring knife　6. boning knife

Part D

Activity Ⅰ

Lady Fingers 手指饼干

Ingredients:原料

4 eggs, separated　4 个鸡蛋,蛋清与蛋黄分离

2/3 cup of white sugar　2/3 杯白糖

7/8 cup of all-purpose flour　7/8 杯中筋面粉

1/2 teaspoon of baking powder　1/2 茶匙发酵粉

Methods:制作方法

1. Preheat oven to 200 ℃. Prepare two 17 inches × 12 inches baking sheets with baking parchment. Fit large pastry bag with a plain 1/2 inch round tube.

预热烤箱到 200 ℃,准备两张 17 英寸×12 英寸的烤盘和烤羊皮纸。用一个普通的 1/2 英寸圆管安装大裱花袋。

2. Place egg whites in the bowl and beat on high speed. Slowly add 2 tablespoons of sugar and

continue beating until stiff and glossy. In another bowl beat egg yolks and remaining sugar. Whip until thick and very pale in color.

将蛋清打入碗里快速搅打，后慢慢加入 2 汤匙白糖，继续搅打直至光滑且能够竖立。在另外一个碗里打入蛋黄和余下的白糖，继续搅打直至浓稠，颜色变白。

3. Sift flour and baking powder together on a sheet of wax paper. Fold half the egg whites into the egg yolks mixture. Fold in flour, and then add the remaining egg whites. Transfer mixture to the pastry bag and pipe out onto prepared baking sheet. Bake 8 minutes.

筛入面粉、发酵粉在蜡纸上，将一半的蛋清加入蛋黄混合物中，加入面粉、余下的蛋清。将混合物装入裱花袋中，挤到之前准备的烤纸上，烤制 8 分钟。

Activity Ⅱ

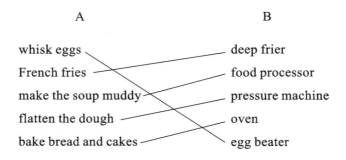

Unit 4　Condiments and Spices

Part A

Activity Ⅰ

1. Tabasco　2. soy sauce　3. wasabi　4. vinaigrette　5. oyster oil
6. fish sauce　7. mustard　8. ketchup　9. honey　10. horse-radish
11. rock sugar　12. vinegar　13. curry　14. maple syrup　15. chili paste

Activity Ⅱ

Variety	Condiments
Saline seasonings	soy sauce, oyster oil, fish sauce
Acid seasonings	vinaigrette, ketchup, vinegar
Hot seasonings	Tabasco, wasabi, mustard, horse-radish, curry, chili paste
Sweet seasonings	honey, rock sugar, maple syrup

Focus on Language　略

Part B

Activity Ⅰ

1. rosemary　2. saffron　3. cinnamon　4. tarragon　5. oregano
6. turmeric　7. peppercorn　8. paprika　9. bay leaf　10. nutmeg
11. fennel　12. clove　13. cumin　14. parsley　15. dill
16. sesame seed　17. basil　18. caraway seed　19. sage　20. marjoram
21. thyme　22. star anise　23. Sichuan pepper　24. mint

Activity Ⅱ

1. Fish gravy is a traditional fermented condiment in coastal areas.

2. Then, you put some condiments and a little salt into the noodles.

3. Sashimi is usually seasoned with wasabi.

4. This meat should be seasoned with salt and mustard.

5. In France, mustard seeds are soaked and then grinded to a paste.

6. Chutney can be mixed with any Indian dish to create a different taste.

7. I would like some ham, sausage, mushrooms, onions, black olives and pineapples for the topping.

8. Salt is a common food preservative.

9. Nutmeg is usually used as flavoring in food.

10. The special test of this soup is due to the saffron powder.

Focus on Language

1. 6 groups   2. root spices, bark spices, leaf spices, flower spices, fruit spices   3. seed spices
4. aromatic   5. bay leaf   6. thyme   7. dill   8. mustard, black pepper   9. garlic, onion, celery
10. ginger

Unit 5   Meat

Part A

Activity Ⅰ

1. 葱爆羊肉   2. 黑椒牛柳   3. 回锅肉   4. 蜜汁肉   5. 清炖狮子头
6. 糖醋里脊   7. 肋眼牛排   8. 丁骨牛排   9. 菲力牛排   10. 侧腹牛排, 牛腩排
11. 牛心   12. 牛腱, 牛腿肉   13. 上脑   14. 西冷牛排   15. 腹腿牛排
16. (牛、羊)后腿部的肉   17. 牛胸肉   18. 肩胛肉

Activity Ⅱ

Twice-cooked Pork

Ingredients:

200 grams of pork rump with skin

Condiments:

100 grams of garlic sprouts	30 grams of soybean paste
9 grams of sweet soybean sauce	1 gram of salt
3 grams of white sugar	1 gram of MSG
3 grams of soy sauce	50 grams of salad oil

Cooking procedures:

1. Put the pork into a wok with cold-water, and then boil it over medium heat for 15 minutes until the pork is well done. Drain the pork and make it cool naturally or freeze it into a refrigerator for 3 to 5 minutes in order to slice it easily.

2. Cut the pork rump into big slices 6 cm×4 cm×0.2 cm and garlic sprouts into chunks.

3. Pour some salad oil in a wok and heat it at the temperature of 150 ℃, and put the sliced pork, salt, Pixian soybean paste, sweet soybean sauce, soy sauce and white sugar. Stir-fry until the dish smells fragrant, next, add the chunked garlic sprouts, MSG and salt for a while, finally dish up.

Focus on Language

1. order   2. steak   3. tenderness   4. time   5. rare   6. flame   7. pink   8. original   9. reduces   10. tough

Translation   略

Part B

Activity Ⅰ

1. chicken 鸡   2. goose 鹅   3. quail 鹌鹑   4. duck 鸭   5. turkey 火鸡   6. squab 乳鸽   7. Beijing roast duck 北京烤鸭   8. baked turkey 烤火鸡   9. spicy duck neck 麻辣鸭脖

Activity Ⅱ

1. pluck   2. cut...segments   3. rinse   4. gut   5. slice   6. stew   7. blanch   8. pour   9. simmer   10. fry   11. squeeze   12. add

Activity Ⅲ   略

Unit 6   Fish

Part A

Activity Ⅰ

1. Spanish mackerel   2. grass carp   3. pomfret   4. ocean perch   5. sardine   6. trout   7. salmon   8. yellow croaker   9. cod   10. tuna   11. catfish   12. crucian

Activity Ⅱ   略

Focus on Language

1. refrigerator   2. preferred   3. longer   4. wrap   5. paper   6. later   7. go bad   8. thaw   9. size   10. refreeze

Translation   略

Part B

Activity Ⅰ

1. shrimp   2. octopus   3. lobster   4. mussel   5. sea snail   6. clam   7. crab   8. oyster   9. scallop

Activity Ⅱ

1. scale   2. cut off   3. gut   4. steam   5. stir-fry   6. bake   7. boil   8. deep-fry   9. toss   10. braise   11. behead   12. devein   13. bone   14. pan-fry   15. smoke

Activity Ⅲ

Fragrant-fried Codfish with Coix Seed

Ingredients：300 grams of codfish   200 grams of coix seed

Accessories：Mussel water   5 grams of basil   5 grams of Italian parsley

Dressing：5 grams of salt   salad oil   5 grams of black pepper

Cooking procedures：

1. Boil coix seed in Mussel water, and cut codfish into small chunks and shred basil.

2. Heat the wok to 300 ℃ and then add a little salad oil, pan-fry codfish for three to four minutes with the codfish skin downwards.

3. Turn the codfish to another side with a cooking shovel, and go on to fry for three to four minutes.

4. Sprinkle with salt, black pepper and olive oil and roast it for seven minutes in the oven at 190 ℃.

5. Boiled coix seed is spread at the bottom of a plate, and sprinkle some Italian parsley and shredded basil onto the surface of it.

6. Spread the cooked codfish over the top of boiled coix seed with plain side onwards, sprinkle salt, decorate with carved and edible flowers.

Part C 略

Unit 7 Fruits and Vegetables

Part A
Activity
1. lemon   2. avocado   3. strawberry   4. orange   5. blueberry   6. lychee   7. lime   8. kiwi fruit   9. raspberry

Focus on Language
1. brown sugar   2. chopped raisins   3. butter   4. boiling water   5. Preheat   6. Remove   7. combine   8. Stuff   9. top with   10. baking pan   11. tender   12. Remove from

Part B
Activity
1. broccoli   2. cauliflower   3. cabbage   4. lettuce   5. purple cabbage   6. asparagus   7. zucchini   8. onion   9. spinach

Focus on Language
1. Lettuce   2. leaves   3. bite size   4. crisp   5. bright   6. dressing   7. plate

Part C
Activity
1. scrub   2. remove   3. wash   4. soak   5. remove the skin of   6. boil   7. poach   8. crush   9. cut open   10. chop

Focus on Language
1. Wash   2. centers   3. boiling water   4. ice-cold   5. peel   6. tomatoes   7. soup

Unit 8 Soup

Part A
Activity
1. consommé of vegetables   2. french onion soup   3. sweet bamboo shoot soup   4. ox-tail soup   5. chicken giblets soup   6. vegetable cream soup   7. fresh tomato cold soup   8. onion soup   9. chicken vegetable soup   10. lobster soup

Part B
Activity
法式洋葱汤

Section one: ingredients

Servings for 10 persons

10 onions

5 tablespoons of butter

7.5 cups of beef broth

7.5 cups of water

2.5 bay leaves

0.25 teaspoon of pepper

0.25 teaspoon of thyme

10 slices of French bread

2.5 cups of Swiss cheese

1 cups of Parmesan cheese

Section two: working process

1. Slice the 10 onions.

2. Cook the onions in butter, over low heat and stir occasionally.

3. After thirty minutes, add the beef broth, water, bay leaves, pepper and thyme, heat and boil them together.

4. Then reduce the heat. Cover and simmer.

5. In the meantime, toast the French bread.

6. After fifteen minutes, put the toasted French bread into bowls and poured onion soup on top.

7. Last sprinkle Swiss and Parmesan cheeses on top.

Part C

Activity

1. Chicken soup.

2. Boned chicken, red date, butter, onion, pepper, lemon juice, egg and salt.

3. Put the chicken into saucepan with the butter, salt, red date, onion and a bit of pepper.

4. A bit of pepper.

5. Yes, of course.

6. I like vegetable soups: clear soup, seaweed soup, tomato and egg soup, and so on.

Part D

Activity

1. Soak dried pears.

2. 8 ounces of shelled green peas.

3. For 2 hours.

4. After 2 hours, boil the ham bone.

5. Yes, skin the eel.

6. Put the eel into pan, with bay leaf, peppercorns, wine and enough water, using high heat to boil. Reduce the heat to low and simmer gently until the eel is tender.

7. It is more delicious.

8. Yes, of course. I had ever done in weekend.

9. Chef.

Unit 9 Pastry

Part A

Activity Ⅰ

1. wonton   2. crispy durian pastry   3. scallion cake
4. steamed stuffed bun   5. fried dough sticks   6. hemp flower
7. cake   8. steamed twisted roll   9. walnut cake

Activity Ⅱ

Elastic dough: ①
Puffy dough: ④⑤⑥⑦⑧
Pastry dough: ②③⑨

Activity Ⅲ

1. What are the characteristics of fried dough sticks?

Fried dough sticks are elongated, yellow and hollow. It smells delicious. It tastes very crispy and very well.

2. Where can buy fried dough sticks?

In China, KFC and McDonald's sell fried dough sticks. Usually, people can also buy them in some breakfast bars.

Focus on Language

1. dumpling   2. vegetables   3. dough   4. pork   5. shrimp   6. cabbage   7. onion   8. eggs
9. traditional   10. festivals

Part B

Activity Ⅰ

1. egg tart   2. pasta   3. birthday cake   4. waffle   5. pizza   6. tiramisu   7. brioche
8. pudding   9. croissant

Activity Ⅱ

1. What materials are usually used to make egg tarts?

Egg tarts are usually made from shortcrust pastry, eggs, sugar, milk or cream, and vanilla, sprinkled with nutmeg and baked.

2. What kind of drink should egg tarts go with?

A cup of Earl Grey tea.

Focus on Language

1. decades   2. Greek   3. specialty   4. tomato   5. cheese   6. leaves   7. delicacy   8. considered
9. Italian   10. favorite

Part C

Activity Ⅰ

1. cake plug-in   2. mounting turntable   3. piping bag
4. nozzle   5. spatula   6. pull wire   7. Korean decoration
8. cream decoration   9. chocolate decoration

Activity Ⅱ

Tools: ①②③④⑤

Types of decoration: ⑥⑦⑧⑨

Focus on Language

1. cream 2. color 3. shade 4. toothpick 5. dip 6. spatula 7. mix 8. amounts
9. intensify 10. mind

Unit 10　Menu Recommendation

Part A

Activity Ⅰ

1. foie gras 2. salad 3. combination 4. tomatoes 5. cucumbers
6. sweet/sour 7. beans 8. rolls 9. tripe 10. prawns
11. oysters 12. beef

Activity Ⅱ　略

Activity Ⅲ

1. Dried fruit: baked almonds/fried peanuts/fried walnuts/potato chips
2. Cold dish: ham/cold sausage/smoked salmon/chicken liver pie/sour cucumber
3. Hot dish: fried prawns/baby hamburger

Part B

Activity Ⅰ

1. steak 2. shrimps 3. lobster 4. chicken 5. beef 6. asparagus 7. pigeon
8. sashimi 9. steamed 10. sea cucumbers 11. duck 12. pig

Activity Ⅱ

1. with 2. with 3. and 4. and 5. in 6. with/in 7. in

Vocabulary

A

a cup of　一杯
a layer of　一层
absorb [əbˈsɔːb]　v. 吸收
accident [ˈæksɪdənt]　n. 事故，意外
accommodation [əˌkɒməˈdeɪʃn]　n. 住处，膳宿
acid [ˈæsɪd]　adj. 酸的
add [æd]　v. 加，加入，增加，添加
all-purpose　通用的
alternate [ɔːlˈtɜːnət]　v. 交替
amino acid [əˈmiːnəʊ ˈæsɪd]　氨基酸
an integral part　不可分割的部分
anise [ˈænɪs]　n. 茴芹
antibacterial [ˌæntibækˈtɪəriəl]　adj. 抗菌的
antiseptic [ˌæntiˈseptɪk]　adj. 防腐的
appetizer [ˈæpɪtaɪzə(r)]　n. 吊胃口的东西，开胃食品
apple [ˈæpl]　n. 苹果
application [ˌæplɪˈkeɪʃn]　n. 应用，申请
appropriate [əˈprəʊpriət]　adj. 合适的
apricot [ˈeɪprɪkɒt]　n. 杏
aroma [əˈrəʊmə]　n. 芳香，香味
aside [əˈsaɪd]　adv. 在旁边
asparagus [əˈspærəgəs]　n. 芦笋
attachment [əˈtætʃmənt]　n. 附件
authentic [ɔːˈθentɪk]　adj. 真正的，真实的
automatic [ˌɔːtəˈmætɪk]　adj. 自动的
avocado [ˌævəˈkɑːdəʊ]　n. 牛油果，鳄梨

B

bacon [ˈbeɪkən]　n. 培根，熏猪肉，咸猪肉，腊肉
bacteria [bækˈtɪəriə]　n. 细菌
baguette [bæˈget]　n. 法棍
bake [beɪk]　v. 烘烤，焙
baker [ˈbeɪkə(r)]　n. 面包师，烘焙师
baking soda　碳酸氢钠，小苏打
ball cutter [bɔːl ˈkʌtə(r)]　挖球器
banana [bəˈnɑːnə]　n. 香蕉
barbecue(BBQ) [ˈbɑːbɪkjuː]　n.（户外烧烤用的）烤架，户外烧烤
　　　　v.（在烤架上）烤，烧烤
basil [ˈbæzl]　n. 罗勒
batch [bætʃ]　n. 一批
batter [ˈbætə(r)]　n. 面糊
bay leaf　月桂叶，香叶
bay [beɪ]　n. 月桂
bean [biːn]　n. 豆，菜豆，豆荚
beat [biːt]　v. 击打
beef brisket　牛胸肉
beef flank　牛肋腹肉
beef round　牛（臀、腿）肉
beef rump　牛后腿部的肉
beetroot [ˈbiːtruːt]　n. 红菜根，甜菜
bell pepper [bel ˈpepə(r)]　甜椒，灯笼椒
bite [baɪt]　v. 咬，叮
　　　　n. 一口食物，少量食物
bivalve [ˈbaɪvælv]　n. 双壳软体动物
black carp　青鱼
black fish　黑鱼
blackberry [ˈblækbəri]　n. 黑莓
blade [bleɪd]　n. 刀片，叶片，桨叶
blanch [blɑːntʃ]　v. 焯（把蔬菜等放在沸水中略微煮一会）
bleach [bliːtʃ]　n. 漂白剂，消毒剂
blend [blend]　v. 混合
blender [ˈblendə(r)]　n. 搅拌器
blister [ˈblɪstə(r)]　n. 水泡
blueberry [ˈbluːbəri]　n. 蓝莓
boil [bɔɪl]　v. 煮
boned [bəʊnd]　adj. 去骨的
boning knife [ˈbəʊnɪŋ naɪf]　拆骨刀，剔肉刀
bowl [bəʊl]　n. 碗

braise [breɪz]　v. 炖，焖
brazier [ˈbreɪziə(r)]　n. 焖锅
breadcrumbs [ˈbredkrʌmz]　n. 面包屑，面包碎
break [breɪk]　v. 打破，打断
breast [brest]　n. （鸡、火鸡的）胸脯肉
brigade [brɪˈɡeɪd]　n. 团队，大军
brioche [briˈɒʃ]　n. 布里欧修
broad bean　蚕豆
broccoli [ˈbrɒkəli]　n. 西蓝花
broil [brɔɪl]　v. 烤，焙（肉或鱼），（使）变得灼热
broiler [ˈbrɔɪlə(r)]　n. 烤箱
broth [brɒθ]　n. （加入蔬菜的）肉汤，鱼汤
brown [braʊn]　n. 棕色
　　　　　　　v. 使变成棕色
　　　　　　　adj. 棕色的
bubble [ˈbʌbl]　n. 气泡
buffet [ˈbʊfeɪ]　n. 自助餐
burn [bɜːn]　v. 烧伤
butcher knife [ˈbʊtʃə(r) naɪf]　屠刀
butcher [ˈbʊtʃə(r)]　n. 屠夫
butter [ˈbʌtə(r)]　n. 黄油，黄油状的食品
　　　　　　　v. 抹黄油于……上，用黄油煎食物，讨好

C

cabbage [ˈkæbɪdʒ]　n. 卷心菜
cake base [beɪs]　蛋糕坯
cake plug-in　蛋糕插件
canned yellow peach　黄桃罐头
cappuccino [ˌkæpuˈtʃiːnəʊ]　n. 卡布奇诺
caraway seed　页蒿子
carbohydrate [ˌkɑːbəʊˈhaɪdreɪt]　n. 碳水化合物
carcass [ˈkɑːkəs]　n. 动物尸体，（尤指供食用的）畜体
carp [kɑːp]　n. 鲤鱼
carrot [ˈkærət]　n. 胡萝卜
catering [ˈkeɪtərɪŋ]　n. （会议或社交活动的）饮食服务，酒席承办
catfish [ˈkætfɪʃ]　n. 鲇鱼
cattle [ˈkætl]　n. 牛
cauliflower [ˈkɒliflaʊə(r)]　n. 花椰菜
celery [ˈseləri]　n. 芹菜

cephalopod [ˈsefələpɒd]　n. 头足动物（如章鱼和乌贼）
cereal [ˈsɪəriəl]　n. 谷类植物，谷物，粮食，谷类食品
cheese [tʃiːz]　n. 奶酪
chef's knife　厨师刀
chef [ʃef]　n. 厨师，主厨，厨师长
cherry [ˈtʃeri]　n. 樱桃
chervil [ˈtʃɜːvɪl]　n. 细叶芹
chicken [ˈtʃɪkɪn]　n. 鸡，鸡肉
Chiffon cake　戚风蛋糕
chili paste　辣酱
chili pepper　红辣椒
chili [ˈtʃɪli]　n. 红辣椒
Chinese cabbage　大白菜
Chinese date [deɪt]　枣子
chives [tʃaɪvz]　n. 细香葱
chocolate decoration　巧克力裱花
cholesterol [kəˈlestərɒl]　n. 胆固醇
chop [tʃɒp]　v. 砍，切
　　　　　　n. 排骨，劈
chopping board　案板，切菜板
chowder [ˈtʃaʊdə(r)]　n. 杂烩
chuck [tʃʌk]　n. 牛肩胛肉，上脑
cilantro [sɪˈlæntrəʊ]　n. 香菜叶
cinnamon [ˈsɪnəmən]　n. 肉桂
clam [klæm]　n. 蛤，蛤蜊
clean as you go　随手清洁
cleaver [ˈkliːvə(r)]　n. 砍肉刀，剁肉刀
clove [kləʊv]　n. 丁香
clump [klʌmp]　v. 使成一丛，使凝结成块
cocktail [ˈkɒkteɪl]　n. 鸡尾酒，开胃食品
cocoa powder　可可粉
coconut [ˈkəʊkənʌt]　n. 椰子
cod [kɒd]　n. 鳕鱼
colander [ˈkʌləndə(r)]　n. 滤锅，漏勺
comb [kəʊm]　n. 蜂巢
commercial [kəˈmɜːʃl]　adj. 商业的
commis [ˈkɒmi]　n. 初级厨师，厨师助理
concentrate [ˈkɒnsntreɪt]　v. 集中，浓缩
condiment [ˈkɒndɪmənt]　n. 调味料，作料
conger eel　大海鳗
connective tissue　结缔组织

consommé [kən'sɒmeɪ] n. 清炖肉汤
container [kən'teɪnə(r)] n. 容器
contaminated [kən'tæmɪneɪtɪd] adj. 受污染的,弄脏的
continue [kən'tɪnjuː] v. 继续,连续
cook [kʊk] n. 厨师
cooking oil 食用油
corkscrew ['kɔːkskruː] n. 瓶塞钻,螺丝锥
corn [kɔːn] n. 玉米
corncob ['kɔːnkɒb] n. 玉米芯
cornstarch ['kɔːnstɑːtʃ] n. 玉米淀粉
cough [kɒf] v. 咳嗽
course [kɔːs] n. 一道菜
cover ['kʌvə(r)] v. 覆盖,代替
crab [kræb] n. 蟹,螃蟹,蟹肉
cranberry ['krænbəri] n. 蔓越莓
cream decoration 奶油裱花
crepe [kreɪp] n. 薄煎饼
crispy durian pastry 榴莲酥
croissant [krwɑːˈsɑ̃ː] n. 牛角面包
cross-contamination [ˌkrɔːskənˌtæmɪ'neɪʃn] n. 交叉污染
crucian 鲫鱼
crush [krʌʃ] v. 压碎
crustacean [krʌ'steɪʃn] n. 甲壳纲动物(如螃蟹、龙虾和虾)
cucumber ['kjuːkʌmbə(r)] n. 黄瓜
cuisine [kwɪ'ziːn] n. 烹饪,风味菜肴
culinary ['kʌlɪnəri] adj. 烹饪的,烹饪用的
cumin ['kʌmɪn] n. 小茴香
curry ['kʌri] n. 咖喱
cylinder ['sɪlɪndə(r)] n. 圆柱体

D

damage ['dæmɪdʒ] v. 损害,损毁
damp cloth 湿布
dangerous ['deɪndʒərəs] adj. 危险的
dark brown sugar 红糖,黑糖
decoration [ˌdekə'reɪʃn] n. 装饰
deep fat fryer 深油炸锅
deep fryer [diːp 'fraɪə(r)] 油炸锅
delicious [dɪ'lɪʃəs] adj. 美味的,可口的
delineate [dɪ'lɪnieɪt] v. 描绘,描写,画……的轮廓
demarara sugar 金砂糖
demerara (产于加勒比地区的)一种红糖
derived from 来源于,源自
designation [ˌdezɪg'neɪʃn] n. 指定
dessert [dɪ'zɜːt] n. 餐后甜食,甜点
dice [daɪs] n. (肉、菜等)丁,小方块
          v. 将(肉、菜等)切成小方块,将……切成丁
diet-conscious ['daɪət'kɒnʃəs] adj. 节食的
dill [dɪl] n. 莳萝
dip [dɪp] v. 蘸
disinfect [ˌdɪsɪn'fekt] v. 给……消毒
do the trick 达到理想的结果,做成功
domestic [də'mestɪk] adj. 本国的,国内的,家用的,家庭的,家务的
double boiler ['dʌbl 'bɔɪlə(r)] 双层蒸锅
dough [dəʊ] n. 生面团
dragon fruit ['drægən fruːt] 火龙果
drain [dreɪn] v. 捞出,沥水
dressing ['dresɪŋ] n. (拌制色拉用的)调料
dried orange peel 陈皮
dried [draɪd] adj. 弄干了的
drumstick ['drʌmstɪk] n. 鼓槌,熟鸡(或家禽)腿下段,下段鸡(或家禽)腿肉
dry [draɪ] adj. 无甜味的
duct [dʌkt] n. 通风道,管道
dumpling ['dʌmplɪŋ] n. 饺子
duplicate ['djuːplɪkeɪt] v. 复制
durian ['dʊəriən] n. 榴莲

E

edible ['edəbl] adj. 可食用的
eel [iːl] n. 鳗鱼
egg beater [eg 'biːtə(r)] 打蛋器
egg tart 蛋挞
egg white 蛋清
egg yolk [jəʊk] 蛋黄
eggplant ['egplɑːnt] n. 茄子
elastic dough 水调面团
electric eggbeater 电动打蛋器
electric [ɪ'lektrɪk] adj. 用电的,电的
eliminate [ɪ'lɪmɪneɪt] v. 消除

endanger [ɪnˈdeɪndʒə(r)]　v. 危及,使遭到危险
endive [ˈendaɪv]　n. 苦苣,菊苣
enhance [ɪnˈhɑːns]　v. 提高,增强
entremets [ˌɒntrəˈmeɪ]　n. 两道菜之间上的一道清淡的菜,甜点
entry-level　入门的,初级的
equal portion　等份
erect [ɪˈrekt]　v. 竖起
essence [ˈesns]　n. 本质,实质,精华,精髓,香精
even [ˈiːvn]　adj. 均匀的,平稳的,相等的,均等的
evenly [ˈiːvnli]　adv. 均匀地
excessive [ɪkˈsesɪv]　adj. 过分的,过度的
executive chef　行政总厨,行政主厨,厨师长
exquisite [ɪkˈskwɪzɪt]　adj. 精致的,细腻的,优美的,剧烈的
extract [ˈekstrækt]　v. 提取,提炼

F

falling down　跌倒
farm-raised　adj. 农场养殖的
fascinating [ˈfæsɪneɪtɪŋ]　adj. 吸引人的
fennel [ˈfenl]　n. 茴香
fermenting box [fərˈmentɪŋ bɒks]　醒发柜
fiber [ˈfaɪbə(r)]　n. 纤维,纤维素
filter [ˈfɪltə(r)]　n. 滤网
fin fish　鱼类
fire extinguisher　灭火器
fish cook　负责烹饪鱼类的厨师
flat [flæt]　adj. 平的
flatfish [ˈflætfɪʃ]　n. 比目鱼
flavor [ˈfleɪvə]　n. 味道,特点,特色
　　　　v. 给……调味,给……增添风味
flip [flɪp]　v. 轻弹,轻击
　　　　n. 跳跃,轻抛
flour [ˈflaʊə(r)]　n. 面粉
foam [fəʊm]　n. 泡沫
　　　　v. 起泡沫
food processor [fuːd ˈprəʊsesə(r)]　食品加工机
fragrance [ˈfreɪɡrəns]　n. 芳香,芬芳,香气,香水
fragrant [ˈfreɪɡrənt]　adj. 芳香的,香的
free-range　adj. 散养的
French fries [frentʃ fraɪz]　炸薯条
frequently [ˈfriːkwəntli]　adv. 频繁地,经常
fried dough sticks　油条
frost [frɒst]　v. 撒上糖霜
fruit [fruːt]　n. 水果
fry cook　负责煎炸的厨师
fry [fraɪ]　v. 油炸,油煎

G

garlic [ˈɡɑːlɪk]　n. 大蒜
garnish [ˈɡɑːnɪʃ]　v. 为(食物)加装饰
　　　　n. (食物上的)装饰菜
gastronomic [ˌɡæstrəˈnɒmɪk]　adj. 烹饪的
gently [ˈdʒentli]　adv. 柔和地,轻轻地
ginger [ˈdʒɪndʒə(r)]　n. 姜,生姜,姜黄色
　　　　v. 使活跃,使有活力
　　　　adj. 姜黄色的
glaze [ɡleɪz]　v. 变得光滑
grain [ɡreɪn]　n. 纹理
grape [ɡreɪp]　n. 葡萄
grapefruit [ˈɡreɪpfruːt]　n. 西柚,葡萄柚
grass carp　草鱼
grate [ɡreɪt]　v. 磨碎,压碎
grater [ˈɡreɪtə(r)]　n. 擦菜板
gravy [ˈɡreɪvi]　n. 肉汁
grease [ɡriːs]　n. 动物油脂,润滑油
　　　　v. 涂油脂于,用油脂润滑
greasy [ˈɡriːsi]　adj. 油腻的,谄媚的,多油的
grouper [ˈɡruːpə(r)]　n. 石斑鱼
gut [ɡʌt]　n. (尤指动物的)内脏,(尤指大的)胃,肚子
　　　　v. 取出……的内脏(以便烹饪)

H

hallmark [ˈhɔːlmɑːk]　n. 特点,品质证明
ham [hæm]　n. 火腿
handful [ˈhændfʊl]　n. 少数,少量,一把(的量)
handle [ˈhændl]　n. 拉手,把手
　　　　v. 应付,应对
hazard [ˈhæzəd]　n. 危害,危险,障碍
head [hed]　n. 头,(植物的茎梗顶端的)头状

叶丛
hemp flower 麻花
herb [hɜːb] n. 香草
hierarchical [ˌhaɪəˈrɑːkɪkl] adj. 分层的,等级体系的
hold [həʊld] v. 拿着,握住
horizontally [ˌhɒrɪˈzɒntəli] adv. 水平地,横向地
hors d'oeuvre [ɔːˈdɜːv] n.(法)(主菜前的)开胃小菜
horse-radish 辣根
hydration [haɪˈdreɪʃn] n. 水化
hygiene [ˈhaɪdʒiːn] n. 卫生

I

ice maker [aɪs ˈmeɪkə(r)] 制冰机
immediately [ɪˈmiːdiətli] adv. 立即,立刻
immigrant [ˈɪmɪɡrənt] n. 移民,侨民,从异地移入的动物(植物)
in most cases 大部分情况下
inch [ɪntʃ] n. 英寸
incorporate [ɪnˈkɔːpəreɪt] v. 合并
infect [ɪnˈfekt] v. 感染,传染
ingredient [ɪnˈɡriːdiənt] n.(混合物的)组成部分,(烹调的)原料,(成功的)要素,因素
injury [ˈɪndʒəri] n. 伤害,损害
intensify [ɪnˈtensɪfaɪ] v.(使)加强
interchangeable [ˌɪntəˈtʃeɪndʒəbl] adj. 可互换的
iron [ˈaɪən] n. 熨斗,铁制品
　　　　　　 v. 熨烫
　　　　　　 adj. 铁质的
issue [ˈɪʃuː] n. 问题

J

juice [dʒuːs] n. 汁,果汁

K

ketchup [ˈketʃəp] n. 番茄酱
kiwi fruit 奇异果
knead [niːd] v. 捏,揉
knob [nɒb] n. 小块
Korean decoration 韩式裱花

L

lard [lɑːd] n. 猪油

larder chef 负责烹饪肉类的厨师长
latte [ˈlɑːteɪ] n. 拿铁
leach out 溶出
leaf [liːf] n. 叶子
leek [liːk] n. 韭葱
leftover [ˈleftəʊvə(r)] n. 吃剩的食物,残羹剩饭
legume [ˈleɡjuːm] n. 豆类
lemon [ˈlemən] n. 柠檬
lettuce [ˈletɪs] n. 生菜
light brown sugar 黄砂糖,黄糖
light cream 淡奶油
lime [laɪm] n. 青柠
limp [lɪmp] adj. 柔软的,不直挺的
liquor [ˈlɪkə(r)] n. 液体,汁,烈性酒
lobster [ˈlɒbstə(r)] n. 龙虾,(供食用的)龙虾肉
ludicrous [ˈluːdɪkrəs] adj. 可笑的,荒唐的
lychee [ˈlaɪtʃiː] n. 荔枝

M

mackerel [ˈmækrəl] n. 鲭(鱼)
maintain [meɪnˈteɪn] v. 维持,继续
management [ˈmænɪdʒmənt] n. 管理,管理人员
mango [ˈmæŋɡəʊ] n. 芒果
maple syrup 枫糖浆
marinade [ˌmærɪˈneɪd] n. 腌泡汁
marjoram [ˈmɑːdʒərəm] n. 墨角兰,牛至
master chef 厨艺大师
meantime [ˈmiːntaɪm] n. 同时,其间
measure [ˈmeʒə(r)] v. 衡量,测量
meat hammer [ˈhæmə(r)] 肉锤
meat [miːt] n. 食用肉类,(猪、牛、羊等的)鲜肉
meatloaf [ˈmiːtləʊf] n. 烘肉卷
medium rare 三分熟
mellow [ˈmeləʊ] adj.(瓜,果等)成熟的,(酒)芳醇的,(颜色或声音)柔和的,老练的
　　　　　　 v.(使)成熟,变柔和,使芳醇
melon [ˈmelən] n. 甜瓜
mince [mɪns] v. 用绞肉机绞(食物,尤指肉)

n. 绞碎的肉，肉末（尤指牛肉）
mineral ['mɪnərəl]　n. 矿物质
mint [mɪnt]　n. 薄荷
mix [mɪks]　v. 混合
mixer ['mɪksə(r)]　n. 搅拌器
mocha ['mɒkə]　n. 摩卡咖啡
molasses [mə'læsɪz]　n. 糖蜜，糖浆
mollusk　n. 软体动物
mount [maʊnt]　v. 安装
mounting turntable ['tɜ:nteɪbl]　裱花转台
mouthwatering ['maʊθwɔ:tərɪŋ]　adj. 令人垂涎的
mozzarella [ˌmɒtsə'relə]　n. 莫泽雷勒干奶酪
multifunctional [ˌmʌlti'fʌŋkʃnl]　adj. 多功能的
muscovado [ˌmʌskə'vɑ:dəʊ]　n. 黑砂糖，粗糖
mushroom ['mʌʃrʊm]　n. 蘑菇
mussel ['mʌsl]　n. 贻贝
mustard ['mʌstəd]　n. 芥末
mysterious [mɪ'stɪəriəs]　adj. 神秘的，不可思议的

N

nozzle ['nɒzl]　n. 裱花嘴
nutmeg ['nʌtmeg]　n. 肉豆蔻
nutrient ['nju:triənt]　n. 营养素
nutritious [nju'trɪʃəs]　adj. 有营养的，营养丰富的

O

obtain [əb'teɪn]　v. 得到，存在
occasionally [ə'keɪʒnəli]　adv. 偶尔，偶然
ocean perch　鲈鱼
octopus ['ɒktəpəs]　n. 章鱼
oily ['ɔɪli]　adj. 油的，油质的
okra ['əʊkrə]　n. 秋葵
olive ['ɒlɪv]　n. 橄榄，油橄榄，橄榄树，橄榄色
adj. 橄榄绿的，黄褐色的，淡褐色的
on the contrary　恰恰相反，正相反，相反地
onion ['ʌnjən]　n. 洋葱
operation [ˌɒpə'reɪʃn]　n. 操作，经营
orange ['ɒrɪndʒ]　n. 橘子，橙子
oregano [ˌɒrɪ'gɑ:nəʊ]　n. 牛至
oven ['ʌvn]　n. 烤箱
overcook [ˌəʊvə'kʊk]　v. 煮得过熟
oyster oil　蚝油
oyster ['ɔɪstə(r)]　n. 牡蛎，蚝

P

paddle ['pædl]　n. 桨叶
palate ['pælət]　n. 腭，上腭，味觉，品尝力
palette knife ['pælət naɪf]　调色刀
pantryman ['pæntrɪmən]　n. 配餐员
paprika [pə'pri:kə]　n. 甜椒粉
paring knife ['peərɪŋ naɪf]　削皮刀，去皮刀
parsley ['pɑ:sli]　n. 荷兰芹，欧芹
partly ['pɑ:tli]　adv. 部分地，不完全地
pasta ['pæstə]　n. 面团（用以制意大利通心粉、细面条等），意大利面食
paste [peɪst]　n. （做点心等用的）加了油脂的面团
pastry chef　面点厨师长
pastry dough　油酥面团
pastry ['peɪstri]　n. 糕点，油酥糕点，油酥面团，油酥面皮
pea [pi:]　n. 豌豆
peach [pi:tʃ]　n. 桃子
pear [peə(r)]　n. 梨子
peel [pi:l]　v. 剥（水果、蔬菜等的）皮，去皮，剥掉
peeler ['pi:lə(r)]　n. 去皮器，削皮器
pepper ['pepə(r)]　n. 甜椒，辣椒，胡椒粉
v. 在……上撒胡椒粉，使布满，连续击打
personal chef　私人厨师
pickle ['pɪkl]　v. 腌渍（泡菜等）
n. 腌菜，泡菜，腌制食品
piece [pi:s]　n. 碎片
pinch [pɪntʃ]　n. 捏，掐，拧
pineapple ['paɪnæpl]　n. 菠萝
piping bag ['paɪpɪŋ bæg]　裱花袋
pitaya ['pɪtəjə]　n. 火龙果
pizza ['pi:tsə]　n. 比萨
plain [pleɪn]　adj. 普通的，不掺杂的
plaster ['plɑ:stə(r)]　v. 抹面
plastic ['plæstɪk]　n. 塑料

adj. 塑料的
plate [pleɪt] n. (烹饪)侧胸腹肉
pluck [plʌk] v. 拔,摘,拔掉(死禽的毛)
plum [plʌm] n. 李子,梅子
poach [pəʊtʃ] v. 水煮(如水煮荷包蛋)
pointed ['pɔɪntɪd] adj. 尖的,尖锐的,严厉的,直截了当的,突出的
policy ['pɒləsi] n. 政策,方针
pomfret ['pɒmfrɪt] n. 鲳鱼
porridge ['pɒrɪdʒ] n. 粥,麦片粥,稀饭
potato [pə'teɪtəʊ] n. 土豆,马铃薯
potentially [pə'tenʃəli] adv. 可能地,潜在地
potman 擦洗锅的人
poultry ['pəʊltri] n. 家禽,禽类的肉
prawn [prɔːn] n. 对虾,大虾
precise [prɪ'saɪs] adj. 明确的
precooked [priː'kʊkt] adj. (食物)煮好的
preheat [ˌpriː'hiːt] v. 预热
prep cook 备餐厨师
prepare [prɪ'peə(r)] v. 准备
prevent [prɪ'vent] v. 预防,防止,阻止
primary ['praɪməri] adj. 主要的
priority [praɪ'ɒrəti] n. 优先,优先权
procedure [prə'siːdʒə(r)] n. 程序,手续,步骤
process ['prəʊses] n. 过程
v. 加工处理
professional [prə'feʃənl] adj. 专业的,职业的
prolific [prə'lɪfɪk] adj. (艺术家、作家等)多产的,众多的,富饶的,(植物、动物等)丰硕的
promptly ['prɒmptli] adv. 迅速地
protein ['prəʊtiːn] n. 蛋白质
prune [pruːn] n. 梅脯,梅干
pudding ['pʊdɪŋ] n. 布丁
puffy dough 膨松面团
pull wire 拉线
pumpkin ['pʌmpkɪn] n. 南瓜
pungent ['pʌndʒənt] adj. 辛辣的,刺激性的,说穿的,一针见血的
puree ['pjʊəreɪ] n. 酱,糊,泥
purify ['pjʊərɪfaɪ] v. 净化,使纯净
purple cabbage 紫甘蓝
purple ['pɜːpl] n. 紫色 adj. 紫色的
purpose ['pɜːpəs] n. 意图,计划

Q

quail [kweɪl] n. 鹌鹑,鹌鹑肉

R

rack [ræk] n. 支架,烧烤架
radix angelicae 白芷
rag [ræɡ] n. 破布
raspberry ['rɑːzbəri] n. 树莓,覆盆子
raw [rɔː] adj. 生的,自然状态的
red date 红枣
red onion [red 'ʌnjən] 红皮洋葱
reduce [rɪ'djuːs] v. 减少,降低
refine [rɪ'faɪn] v. 精炼,提纯
refrigerate [rɪ'frɪdʒəreɪt] v. 冷藏,冷冻
refrigeration [rɪˌfrɪdʒə'reɪʃn] n. 冷冻,制冷
regulation [ˌreɡju'leɪʃn] n. 管理,规则
reheat [ˌriː'hiːt] v. 重新加热
relief cook 替班厨师
remaining [rɪ'meɪnɪŋ] adj. 剩下的
resolve [rɪ'zɒlv] v. 解决
retain [rɪ'teɪn] v. 保持
rib [rɪb] n. 肋骨,排骨
ribbon fish 带鱼
ribbon ['rɪbən] n. 带,缎带,带状物
v. 把……撕成条带
rice flour dough 米粉面团
rinse [rɪns] v. (用清水)冲洗,洗刷
n. 冲洗
risotto [rɪ'zɒtəʊ] n. 意大利肉汁烩饭
roast [rəʊst] v. 烤
n. 烤肉
rock sugar 冰糖
roll up 卷起
rolling pin ['rəʊlɪŋ pɪn] 擀面杖
rosemary ['rəʊzməri] n. 迷迭香
rotate [rəʊ'teɪt] v. (使)旋转
rub [rʌb] v. 擦,搓
rubber ['rʌbə(r)] n. 橡胶,橡皮擦
adj. 橡胶做的

S

saffron ['sæfrən] n. 藏红花
sage [seɪdʒ] n. 鼠尾草
salad ['sæləd] n. 沙拉,色拉

saline [ˈseɪlaɪn]　adj. 含盐的，咸的
salmon [ˈsæmən]　n. 鲑，三文鱼
salt [sɔːlt]　n. 盐
salted [ˈsɔːltɪd]　v. 在食物中放盐，用盐腌制
sanitation [ˌsænɪˈteɪʃn]　n. 公共卫生
sardine [ˌsɑːˈdiːn]　n. 沙丁鱼
sauce chef　调汁厨师
saucepan [ˈsɔːspən]　n. 炖锅
sausage [ˈsɒsɪdʒ]　n. 香肠，腊肠
sauté [ˈsəʊteɪ]　v. 炒，煸，嫩煎
savory [ˈseɪvərɪ]　n. (烹调用的)香薄荷
　　　　adj. 好吃的，咸味的
scald [skɔːld]　v. (不加任何调料地煮熟)白灼
scale [skeɪl]　v. 去鳞
scallion [ˈskælɪən]　n. 葱
scallop [ˈskɒləp]　n. 扇贝
scrape [skreɪp]　v. 刮，擦
scratchy [ˈskrætʃɪ]　adj. 刺痛的，发痒的
scrub [skrʌb]　v. 擦洗
seafood [ˈsiːfuːd]　n. 海鲜
sear [sɪə(r)]　v. 烧焦
season [ˈsiːzn]　n. 季节
　　　　v. 调味
seasoning [ˈsiːzənɪŋ]　n. 调味品
second in command　副指挥，第二把手
serrated knife [seˈreɪtɪd naɪf]　锯齿刀
sesame seed　芝麻，芝麻籽
sesame [ˈsesəmɪ]　n. 芝麻
severe [sɪˈvɪə(r)]　adj. 严重的
shade [ʃeɪd]　n. 色度
shallot [ʃəˈlɒt]　n. 葱，红葱头
shank [ʃæŋk]　n. (烹饪)牛腱，牛腿肉
shellfish [ˈʃelfɪʃ]
shred [ʃred]　v. 将(肉、菜等)切成丝，切碎
　　　　n. (撕或切的)细条，碎片
shrimp [ʃrɪmp]　n. 虾，小虾
Sichuan pepper　花椒
sieve [sɪv]　n. 筛子，滤网
　　　　v. 筛，筛选，滤
sift [sɪft]　v. 筛选，滤
silver carp　鲢鱼
simmer [ˈsɪmə(r)]　n. & v. 炖，慢煮
sirloin [ˈsɜːlɔɪn]　n. 牛里脊肉，牛上腰肉

sit [sɪt]　v. 坐，放置
skeleton [ˈskelɪtn]　n. 骨骼，骨架
skimmer [ˈskɪmə(r)]　n. 撇取者，撇取物
skin [skɪn]　v. 去皮，剥去……的皮
sleeve [sliːv]　n. 袖子
slice up　将……切成薄片
slice [slaɪs]　n. (切下的食物)薄片，片
　　　　v. 把……切成(薄)片，切
slicer [ˈslaɪsə]　n. 切片机
slide [slaɪd]　v. 滑动
slightly [ˈslaɪtlɪ]　adv. 轻微地
slip [slɪp]　v. 滑动，滑倒
slippery [ˈslɪpərɪ]　adj. 滑的
slotted [ˈslɒtɪd]　adj. 开槽的
smoke [sməʊk]　n. & v. 烟，熏
smoked salmon [sməʊkt ˈsæmən]　烟熏三文鱼
smooth [smuːð]　adj. 光滑的，调匀的
snag [snæg]　v. 被绊住，形成障碍
sneeze [sniːz]　v. 打喷嚏
soak [səʊk]　v. 浸泡
soapy water　肥皂水
soft [sɒft]　adj. 软的
sommelier [ˈsɒməljeɪ]　n. (法)酒侍，品酒师
soothe [suːð]　v. 安慰，安抚
sour cream　酸奶油
sous chef　副厨师长，副主厨
soybean paste　黄酱
soy sauce　酱油
soybean [ˈsɔɪbiːn]　n. 大豆，黄豆
Spanish mackerel　鲅鱼
spatula [ˈspætʃələ]　n. (搅拌或涂敷用的)铲，抹刀，刮刀
specialty [ˈspeʃəltɪ]　n. 特色，专长
spice [spaɪs]　n. 香料
spicy [ˈspaɪsɪ]　adj. 辛辣的，加有香料的
spill [spɪl]　v. 溅出，洒出
spoon [spuːn]　n. 勺，匙，调羹，一匙的量，匙状物，匙桨
　　　　v. 用汤匙舀取，轻轻向上击
spread [spred]　v. 展开，伸开，传播，涂，分摊
　　　　n. 散布，广泛
sprig [sprɪg]　n. (烹饪或装饰用)小枝
sprinkle [ˈsprɪŋkl]　v. 撒，洒

squab [skwɒb]　n. 乳鸽
squeeze [skwiːz]　v. 挤压，压榨
squid [skwɪd]　n. 乌贼，鱿鱼
stack [stæk]　n.（整齐的）一堆
star anise　八角茴香
starch [stɑːtʃ]　n. 淀粉，含淀粉的食物
steak [steɪk]　n. 牛排，肉排
steam [stiːm]　n. 水蒸气
steel [stiːl]　n. 钢，钢铁，钢制品
sterilize [ˈsterəlaɪz]　v. 消毒，杀菌
stiff [stɪf]　adj. 严厉的，僵硬的
stir [stɜː(r)]　v. 搅拌
stir-fry [ˈstɜːfraɪ]　v. 用旺火炒
　　　　　　　　n. 炒菜
stock [stɒk]　n. 现货，存货，库存；（尤指烹饪）高汤
stockpot [ˈstɒkpɒt]　n. 汤锅
strawberry [ˈstrɔːbəri]　n. 草莓
stray [streɪ]　adj. 走失的，孤立的
streaky pork [ˈstriːki pɔːk]　五花肉
string bean [ˌstrɪŋ ˈbiːn]　菜豆
strip [strɪp]　v.（烹饪）把……切成条状
　　　　　　n.（纸、金属、织物等）条，带
strong-flavored　味道浓烈的
stuffing　n. 填充物，填料
substantial [səbˈstænʃl]　adj. 大量的
sugar [ˈʃʊɡə(r)]　n. 糖
supervise [ˈsuːpəvaɪz]　v. 监督
swirl [swɜːl]　v. 旋转
swiveling [ˈswɪvlɪŋ]　adj. 转动的
symptom [ˈsɪmptəm]　n. 症状

T

Tabasco [təˈbæskəʊ]　n.（塔巴斯科）辣椒酱
take precautions　采取预防措施
tarragon [ˈtærəɡən]　n. 龙蒿
technology [tekˈnɒlədʒi]　n. 科技，工艺
tender [ˈtendə(r)]　adj. 嫩的，柔软的
tenderloin [ˈtendəlɔɪn]　n.（牛、猪、羊的）里脊肉，嫩腰肉
thaw [θɔː]　v.（结冰后）解冻，融化
thicken [ˈθɪkən]　v. 使变厚（或粗、密、浓）
thigh [θaɪ]　n. 大腿，股，食用的鸡（等的）大腿

thoroughly [ˈθʌrəli]　adv. 非常，极其，彻底，完全
thyme [taɪm]　n. 百里香
tiny [ˈtaɪni]　adj. 极小的
tiramisu [ˌtɪrəˈmiːsuː]　n. 提拉米苏
tissue [ˈtɪʃuː]　n.（人、动植物的）组织，（尤指用作手帕的）纸巾
toast [təʊst]　v. 烘，爆
toaster [ˈtəʊstə(r)]　n. 烤面包机
tomato [təˈmɑːtəʊ]　n. 西红柿，番茄
tongs [tɒŋz]　n. V 形夹子
toothpick [ˈtuːθpɪk]　n. 牙签
trip [trɪp]　v. 绊倒
trout [traʊt]　n. 鳟鱼
tube [tjuːb]　n. 管子，管状物
tuna [ˈtjuːnə]　n. 金枪鱼
turkey [ˈtɜːki]　n. 火鸡
turmeric [ˈtɜːmərɪk]　n. 姜黄，姜黄根粉
turnip [ˈtɜːnɪp]　n. 芜菁

U

ulcer [ˈʌlsə(r)]　n. 溃疡
uniform [ˈjuːnɪfɔːm]　n. 制服
uniformly [ˈjuːnɪfɔːmli]　adv. 均匀地，一致地
univalve　n. 单壳软体动物
upper [ˈʌpə(r)]　adj. 较高的
utensil [juːˈtensl]　n. 餐具，炊具

V

valuable [ˈvæljuəbl]　adj. 有价值的，宝贵的
vanilla [vəˈnɪlə]　n. 香草香精
variegated carp　鳊鱼
vegetable chef　负责烹饪蔬菜的厨师长
vegetable [ˈvedʒtəbl]　n. 蔬菜
vegetarian [ˌvedʒəˈteəriən]　n. 素食者，食草动物
　　　　　　　　　　　　adj. 素食者的，素菜的
vent [vent]　n. 进出口，通风口
versatile [ˈvɜːsətaɪl]　adj. 多用途的，多才多艺的
versatility　n. 用途广泛，技术全面
vinaigrette [ˌvɪnɪˈɡret]　n. 油醋汁
vinegar [ˈvɪnɪɡə(r)]　n. 醋

vitamin ['vɪtəmɪn]　n. 维生素，维他命

W

waffle ['wɒfl]　n. 华夫饼
walnut cake　桃酥
walnut ['wɔːlnʌt]　n. 核桃
wasabi　青芥末
watercress ['wɔːtəkres]　n. 西洋菜
watermelon ['wɔːtəmelən]　n. 西瓜
wheel [wiːl]　n. 轮子，旋转
whisk [wɪsk]　n. 搅拌器
　　　　　　　v. 搅拌
white wine　白葡萄酒
wine [waɪn]　n. 葡萄酒，深红色
winter squash [skwɒʃ]　节瓜
wipe up　擦干净，擦掉
wipe [waɪp]　v. 抹
wok [wɒk]　n. 炒菜锅
wonton [ˌwɒn'tɒn]　n. 馄饨
work [wɜːk]　v. 揉（面团），搅（黄油）
wrap [ræp]　v. 包起来

Y

yellow croaker　黄花鱼

Z

zucchini [zuˈkiːni]　n. 西葫芦
8-tooth [tuːθ]　8齿